Power Up Your Classroom With AI

Explore The Origins and Ethics of AI
Craft Engaging AI Activities
Streamline Admin Duties & Lesson Planning

pjs

PJS Publishing LLC
Post Box 230663
Las Vegas, NV 89105

Published in the United States of America
Author: Peter James Shankland
Printed by: Amazon/KDP

CONTENTS

WELCOME TO POWER UP YOUR CLASSROOM WITH AI!

My name is Peter J Shankland and I am deeply grateful that you have selected this book for use in bringing AI (artificial intelligence) into your classroom. Having taught for 30 years, I am excited to share with you the essential knowledge necessary to embrace AI use as you seek to successfully integrate this game-changing technology into your lesson plans in pursuit of K-12 excellence.

AI's footprint in educational environments has expanded rapidly. Reports indicate that AI use in U.S. classrooms has surged by over 50% since 2020. As such, it is paramount that educators gain fluency in AI tools and usage.

According to the Department of Education (*EdTech* magazine),

> *"The rapid growth of AI in education is reshaping the way we teach and learn, offering personalized learning experiences, automating routine tasks, and supporting educators. As AI continues to advance, it holds the potential to revolutionize classrooms by providing tools that cater to indi-*

vidual student needs and freeing up teachers to focus on more meaningful interactions with students."

The proverbial train has left the station so let's hop on board.

Power UP Your Classroom With AI has two (2) parts.

Part One lays out foundational knowledge starting with:

- AI origins - because knowing and sharing context with students is important,
- Ethical concerns and the anxiety arising from AI use,
- Integration of AI into lesson plans; and,
- Resources for professional development and community building.

Part Two turns our attention to the hands-on aspect of AI in education:

- Resource links by subject and grade level,
- Effective AI use for activity creation,
- Sample activities to inspire; and,
- A brief look towards our AI future.

Over 150+ resource links are shared throughout the text.

At its' core, **Power UP Your Classroom With AI** seeks to provide the knowledge and skills necessary to transform any AI hesitancy you may have into confidence - to address the here and now of blending this new technology with the best aspects of traditional teaching. **Enjoy the journey!**

PART I

1

AI ORIGINS AND THE HOLY GRAIL

Let's begin by exploring the origins of AI and the significant technological advancements that have created today's exciting, educational environment. With historical understanding comes an appreciation for the AI tools now available.

Moreover, sharing such an understanding with your students will provide the critical context they need to appreciate the changes they are encountering in your classroom and their daily lives.

1.1 THE BIRTH OF AI: FROM TURING TO TODAY

In his groundbreaking 1950 paper, *Computing Machinery and Intelligence,* British professor Alan Turing proposed, in colloquial terms, the *Turing Test.* He posited that a machine's ability to exhibit intelligent behavior is indistinguishable from a human. In the test, the tester interacts blindly with both a human and a machine through a computer interface.

If he/she cannot reliably tell which is which, the machine is considered to have passed the test - demonstrating human-like intelligence. While the test did not assess the machine's actual understanding or consciousness, its ability to effectively mimic human responses was groundbreaking.

Since then, the field of AI has moved from mere theory to the first computers that performed straightforward calculations, to the complex, interactive machines they are today.

On the coattails of Turing, several other luminaries gained prominence.

Marvin Minsky, commonly called the *Father of AI,* co-founded the Massachusetts Institute of Technology's AI laboratory in 1959. His work on the *Perceptron,* an early *neural network,* laid the foundation for many of the AI applications we use today. The following analogy brings the *neural network* concept to life.

Imagine teaching a child to differentiate between various fruits. You show them apples, bananas, and oranges, each time pointing out the distinctive features. Eventually, the child learns to recognize the fruit based on these characteristics without your help.

This is essentially what machine learning algorithms do - they grasp concepts from the repetitive input of data and learn. They create neural networks, akin to the neurons and synaptic pathways that comprise our brain.

John McCarthy, who coined the term *Artificial Intelligence* in 1956, was instrumental in developing programming languages like *Lisp* which gave AI programming the ability to handle symbolic information effectively. In brief, *Lisp* created powerful abstractions from data leading to advanced computations.

Geoffrey Hinton, a British/Canadian cognitive psychologist and computer scientist, is prominent for his work in the 80's and 90's on *back propagation* and neural networks. These allow computer networks to adjust their parameters to minimize error and improve a computer's learning accuracy. *Back propagation identifies a computer's capacity to uptake new information, reflect on it, then reconsider and re-align the new with the old, subsequently changing and updating it as required.*

The creation of neural networks and the exponential increase in computational power have propelled AI advancements at breakneck speeds and allow educators to leverage these technologies in the classroom to enhance teaching and learning experiences.

1.2 IMPACT ON EDUCATIONAL SOFTWARE

Neural networks - smart computer systems that learn from data – are providing exponentially larger amounts of information and are changing the way we teach and learn such that today's educational tools are smarter and more helpful than ever before.

For example, better student assessments. Traditional tests sometimes miss the full picture of what a student knows or misunderstands. But neural networks can look at data from quizzes, interactive lessons, and even games to give a clearer view of a student's true understanding - not just what they've memorized.

Because of this, educational software is becoming more personalized, engaging, and effective. It's helping students learn in ways that suit their pace and style, while giving teachers the freedom to do what they do best: inspire, connect, and guide their students.

In a world where every student learns differently, relying on one-size-fits-all teaching is outdated. Neural networks are leading the way to a more flexible, student-centered future. **This really is the Holy Grail of AI and education.**

Keeping this vision in mind, please know that **Power UP**'s primary goal is fulfilling the educational needs of the here and now. This means:

- resources for streamlining administrative duties,
- step-by-step AI integration at a pace that cultivates confidence,
- sharing essential AI knowledge for effective use of this powerful technology,
- hands-on activities necessary to ensure success in your classroom; and more!

Let's work together in building a 21st century education that creates the causes and conditions for all students to excel.

1.3 AI IN SELF-DIRECTED LEARNING CLASSROOMS

In a self-directed learning environment, AI tools become thoughtful partners. They adapt content, provide feedback, and provide insights into how students learn best. **But it's teachers who make the magic happen** - bringing creativity, empathy, and connection that no algorithm can replicate.

AI helps students take charge of their learning journey. **Teachers help students make sense and give meaning to it**. Together, you create classrooms that are personalized, engaging, and deeply human.

An aside: *Before speaking directly to the issue of personalized learning, please allow me to share a highly impactful story from my childhood that relates directly to this topic.*

My mother was an elementary and junior high teacher for 40 years. In 1974, prior to my entering 6th grade, she heard about an extraordinary teacher (Ms. B) doing ground-breaking things in her classroom at the school I attended (Mom worked elsewhere). She petitioned the principal to ensure that I entered this dynamic educator's class. I am so grateful she did. Here's what was so special.

Ms. B filled the classroom with plants, terrariums, and various, what we now call STEM based activities that ignited interest and wonder. Wherever you looked, there were cool things to do and learn. She had us bring in a large cardboard box that we cut and placed around our desks to make a personal cubicle about 4 feet high that we decorated as we wished.

Then, on entering the classroom each day, we found the surrounding chalkboards filled with all the activities we had to accomplish that day - individually or in small groups. Our choice. She would monitor our work individually or sometimes gather our attention and teach the entire class.

We felt empowered and in control of our learning at our own pace. Later, as an adult, I realized she had been in control the entire time. I share this story because, in my mind, this was an early, computer free version of the personalized learning AI realizes.

Today, imagine your classroom functioning in a similar fashion. Only this time, it is empowered and driven by AI adaptive learning platforms where each student receives personalized AI attention, tailored to their learning pace and style.

AI adaptive lessons in Mathematics allow AI to analyze student responses in real time and adjust the difficulty, type, and pace of questions accordingly. This means if a student excels at fractions but struggles with decimals, the system will adjust to provide more support where needed.

Likewise, language learning tools use sophisticated AI algorithms to determine with which words and phrases users struggle and tailor future lessons to address those weaknesses.

Such personalization makes learning more effective and enhances engagement because the content is directly relevant to their needs. Imagine the shifts in your teaching approach and subsequent behaviors – how you can more effectively use your time and cater to specific student needs.

Having said all that, such a paradigm shift requires change at all levels - from the classroom to state governments. We, as educators, must re-imagine ourselves as guides, rather than acting within the role of traditional teachers.

Parents must be aware and on board.

Principals must create a healthy, supportive environment.

School boards must embrace and encourage change at every level.

Lastly, the powers that be must provide appropriate funding for the range of infrastructure changes required to realize this vision. All of this can and will happen if, and when, we act in concert.

That being said, we shift focus to an unfolding issue that is critical not only to education, but our society at large - AI Ethics.

2

AI ETHICS

2.1 DEFINING ETHICAL AI USE

Ethical AI use in educational settings sees AI as a tool that encourages positive educational outcomes while respecting the dignity of all students. Ethical AI use means the technologies are employed in a fair, transparent, and accountable manner. It means that the tools being used promote and satisfy a range of ethical concerns.

The Institute for Ethical AI in Education (**https://www.ai-in-education.co.uk/**) has drafted guidelines to ensure AI applications are transparent and understandable to users. Critically, **they cast AI tools as complementary to educators' efforts, rather than serving as the primary drivers of education.**

As an educator, it is critical that ethical AI use underpins everything done with this new technology. Tools must be vetted properly for both their educational value and their ethical implications. You must decide which AI tools enter your classroom and how they are used. As you research software, ask yourself:

- Is it inclusive?
- Is the AI tool free from biases that could harm students?
- Is it transparent and clear in what it aims to accomplish?
- How does the AI service provider protect student data?

These are all very critical questions. Surf around. Explore sites. Experiment with AI tools. Reach out to creators and ask questions.

By staying aware and informed during implementation, you can safeguard your students' rights and foster a trust-based learning environment where technology serves as a bridge to knowledge, not a barrier.

2.2 FOSTERING AN ETHICAL CLASSROOM CULTURE

Creating an ethical classroom culture is much more than simply using the right tools; it's about cultivating the right mindset among your students. Regularly discuss with students the AI tools you are using. Explain how they work and discuss their potential benefits and limitations.

Raise the issue of writing papers, completing homework or doing just about any classroom exercise with AI and claiming it as your own work.

Establish guidelines for proper behavior when on-line. This transparency prepares students to think about, and use, AI ethically.

Here are ten (10) suggested guidelines for ethical AI use when doing schoolwork. Consider posting them for ease of reference and as a constant reminder.

a. **Use AI as a Tool, Not a Shortcut**: AI can assist with research, ideas, and checking work, but do not use it to write assignments or complete tasks for you. Always put in your own effort and learning.
b. **Cite Your Sources**: When using information from the internet or AI tools, always give proper credit to the original creators by citing sources and avoiding plagiarism.
c. **Verify Information**: Not everything on the internet or AI-generated content is accurate. Cross-check facts using trusted educational sources to ensure the information is reliable.
d. **Respect Privacy**: Do not share your personal information or the personal information of others when using AI platforms, apps, or websites. Maintain privacy and stay safe online.
e. **AI Responsibly**: AI tools are powerful but should be used with responsibility. Avoid using them to cheat on assignments or exams. Instead, use them to support your learning process.
f. **Avoid Inappropriate Content**: When researching online, stick to educational, age-appropriate websites and tools. Avoid content that is irrelevant or unsuitable for your academic level.
g. **Think Critically**: Always question the content you find online or through AI tools. Analyze, interpret, and reflect on how it aligns with what you are learning in school.
h. **Don't Misrepresent Your Work**: Submitting AI-generated or copied content as your own violates academic honesty. Always be clear about what you've done versus what was simply and solely AI generated.
i. **Stay Within Assignment Guidelines**: Follow your teacher's instructions regarding how AI and online

resources should be used. If the use of AI is restricted or requires permission, respect those guidelines.

j. **Practice Digital Etiquette**: Engage politely and professionally when using online discussion boards, forums, or group projects. Maintain respectful communication with teachers and peers when collaborating online.

2.3 ENCOURAGING ETHICAL THINKING

a. **Critical Thinking Exercises:** Conduct exercises that encourage students to consider AI's ethical implications for the specific matter at hand. For example: present a scenario where an AI tool decides a student's grade. Ask your students to analyze the fairness and implications of this scenario.
b. **Diverse Perspectives:** Encourage students to consider how AI impacts different people, differently. This could involve discussions about AI and accessibility or how cultural biases can be embedded in algorithms.
c. **Role-Playing:** For older students, ask them to imagine themselves as the various stakeholders in the AI development process, such as programmers, users, and ethical auditors. This activity can help students understand the diverse considerations that must be balanced when developing and utilizing AI technologies.
d. **Ethical Decision-Making Frameworks:** Teach students to apply ethical decision-making frameworks to AI scenarios. This could be as simple as asking, "*Who benefits from this AI tool, and who might be disadvantaged?*" Such questions prompt students to think beyond the immediate utility of technology.

By integrating these practices into your classroom, you are educating students about AI and empowering them to confidently navigate its' complex ethical landscape. Together with the academic benefits, this proactive approach prepares them to be thoughtful, informed citizens in our ever-increasing, technologically advanced world.

Further, fostering an ethical classroom culture expands your role beyond the curriculum. You are shaping how future generations will think about technology and its place in society. It's about making ethical considerations as reflexive as checking for spelling errors or verifying sources in a research paper.

In a classroom infused with such an ethical mindset, AI becomes a tool for learning about technology and ethics alike; this prepares students for both schoolwork and the ethical challenges of tomorrow.

2.4 CREATING A RESPONSIBLE AI POLICY

Together with the student specific guidelines above, your navigation of this new learning is best served by advocating for a school-wide AI policy that aligns with your institution's educational objectives and goals. Make ethical guidelines the backbone of your approach in defining how AI should, and should not, be used.

Emphasize respect for the range of ethical issues this chapter explores.

A well-crafted AI policy:

- outlines clear objectives - What exactly is the desired outcome when integrating a particular AI tool? What is/are the goal(s)?

- creates a framework for an effective policy based on these goals; and,
- defines the roles and responsibilities of everyone involved - you, students, school IT staff, and parents. Who is responsible for AI infrastructure? Who monitors the data privacy and security? What are healthy student behaviors? Definition and clarity help avoid any confusion.

Lastly, a well-crafted AI use policy establishes clear enforcement mechanisms. It's crucial to outline what happens if the policy is breached. This could range from a simple review and discussion to more concrete steps if privacy or security are compromised.

2.5 COLLABORATIVE POLICY DEVELOPMENT

AI policy creation is a collaborative effort. Engage with fellow teachers, school administrators, and IT staff to gather varied perspectives and expertise. Include students in the conversation so they feel invested in responsible AI use.

Parents can also be invaluable collaborators. They might have perspectives or concerns that had not crossed your mind, especially when it comes to data privacy and security. Hosting a workshop or a meeting where parents can voice their thoughts and learn about the AI technologies being used helps build a community of trust around the AI initiatives at your school.

2.6 ALIGN AI POLICY WITH SCHOOL VALUES

The best AI policies support your school's broader educational goals and values. This is crucial for success and acceptance. Begin by reviewing your school's mission statement and its learning objectives. In what ways can AI integration aid in realizing these objectives?

For example, if a school charter values inclusivity, ensure that the AI tools enacted do not unintentionally exclude certain groups of students through biased algorithms or inaccessible technology.

Consult your school's current technology policies and see if you can build on the existing framework, or, decide whether the policies need to be updated in-light of new AI advances. Such practices promote consistency and alignment across all areas of AI use. This both simplifies tool management and strengthens the overall educational strategy.

AI is a rapidly evolving field. Therefore, ongoing education about its advances and ethical considerations empowers teachers to use AI more effectively and responsibly. Aligning policies with professional development opportunities fosters a culture of continuous learning and adaptation, which is essential in the digital age.

Throughout this policy-crafting process, always remember that you're not just imparting knowledge; you're shaping how young minds perceive and interact with the technologies. A well-thought-out AI policy protects and guides; it educates and empowers. It makes certain that AI tools are used effectively, wisely, and ethically.

2.7 NAVIGATING PRIVACY CONCERNS WITH STUDENT DATA

Understanding and managing data privacy is crucial for successful AI integration. The legal framework in the United States, primarily governed by the Family Educational Rights and Privacy Act (FERPA), establishes how student information is to be handled.

FERPA requires that schools have written consent from parents or eligible students to release any information pertaining to the educational record. When AI tools are used, however, the lines can sometimes blur on what is considered part of an *"education rec ord"* and who has rightful access.

Some AI educational tools, for example, might analyze student performance data to adjust teaching methods; but, if this data is stored or processed by third-party AI providers, it could create privacy concerns if not properly managed.

Further, the implications of AI use extend beyond simple compliance with laws like FERPA. Sure, FERPA defines legal standards to help avoid legal repercussions. But, the larger picture is about striking a healthy, trusting balance between in both educational benefits and related privacy concerns.

2.8 BEST PRACTICES FOR DATA MANAGEMENT

Managing student data with AI tools requires a commitment to best practices. Here are some guidelines to help you manage student data responsibly.

2.8.1) Data Minimization: Only collect necessary data. It's tempting to gather as much information as possible to improve learning outcomes, but more data can mean greater risk if a breach occurs.

2.8.2) Secure Storage Practices: Ensure that any data collected is stored securely, using encryption and other security measures. Also, verify that any third-party AI providers you may use adhere to the same high standards of data security.

2.8.3) Regular Audits: Conduct regular audits of your AI tools and data management practices. This helps identify any potential vulnerabilities and ensures compliance with both legal standards and your school's policies.

2.8.4) Clear Data Usage Policies: Have clear policies in place that outline how student data is used, who has access to it, and under what circumstances. This transparency is key to maintaining trust.

2.8.5) Ongoing Training: Regularly train staff on the latest best practices for data security and privacy. As new threats emerge and technologies evolve, staying informed is your best defense.

2.9 ENGAGING WITH PARENTS AND GUARDIANS

Transparent communication with parents and guardians about how AI tools use student data is crucial. Here's how you can engender their support and trust:

2.9.1) Information Sessions: Hold regular information sessions to educate parents about the AI tools being used, what data is being collected, and how it is being protected.

2.9.2) Feedback Opportunities: Provide opportunities for parents to give feedback on the use of AI tools in your classroom. Three (3) options include surveys, comment boxes, or parent-teacher meetings.

2.9.3) Clear Communication: Use clear, jargon-free language in all communications about AI and data privacy. Ensure that all parents, regardless of their tech-savviness level, can understand how their children's data is being used and protected.

2.9.4) Opt-Out Options: Whenever possible, provide options for parents to opt out of data collection by AI tools. This respects their preferences and reinforces trust.

Actively engaging with parents and making them partners in AI use alleviates their concerns. It demonstrates that you are using the AI tools responsibly and effectively. This partnership is integral in creating an educational environment where technology serves as a bridge to better learning outcomes and is grounded in trust and mutual respect.

2.10 FOUR (4) CONCERNS REGARDING AI CONTENT CREATION

Bringing AI into our classrooms is akin to hiring a new teaching assistant. Therefore, just as we require an interview for a human assistant, we must ensure our digital assistant undergoes proper vetting as well.

This means being vigilant about a) biases in AI algorithms, b) advocating for inclusive AI design, c) demanding equitable access to AI resources, and d) requiring cultural competence in all AI applications.

2.10.1) Biases in AI Algorithms

Imagine you're using an AI tool that helps tailor math problems to each student's learning level. It sounds fantastic, right? But, what if the data used to train the AI tool had mostly come from affluent suburban schools with abundant resources, and you're teaching in a diverse, under-funded school?

There is a real risk that AI might not perform as well for the latter students because it wasn't properly trained - meaning, the data that allows a program to learn - is incomplete. In creation, it failed to encompass a broad, sufficiently diverse dataset. This is just one example of how bias in AI algorithms can sneak in and potentially lead to biased practices.

It's crucial to understand that AI isn't just about handling data - it's about handling data responsibly. Look into whether the AI tools you're considering have been tested across diverse educational settings. Reach out to the tool creators and ask tough questions:

- Where did the data come from?
- Exactly what data was used?
- Who does it represent?
- Who does it leave out?

Finding acceptable answers to these questions means that the AI tools used in your classroom offer equal opportunities for learning to all students, regardless of their background.

AI biases can subtly influence student outcomes in ways not immediately apparent. For instance, an AI program designed to recommend courses to students might inadvertently steer them toward subjects that align with stereotypical notions of what is appropriate for their gender, race, or socioeconomic status, rather than a student's personal interests and abilities.

Understanding the origins and impacts of these biases is the first step in countering them. It helps us be proactive in ensuring the AI tools used serve all students fairly and equitably. Awareness alone, however, isn't enough.

We need actionable strategies to effectively mitigate these biases. Mitigating bias in AI tools is akin to tending a garden. It requires regular attention and care.

Here are some suggested strategies to address this issue:

- **Involve Diverse Teams When Selecting AI Tools:** Teams that create, purchase, and/or manage AI tools should represent multiple perspectives and backgrounds. Varied insights ensure that the AI tools are scrutinized through a broader lens, catching subtleties that might otherwise go unnoticed.

- **Implement Bias Audits:** Regularly auditing AI tools for bias is akin to routine check-ups. These audits can be conducted by internal teams or external experts and should involve testing the AI's decisions to ensure they are fair and unbiased. If biases are detected, the algorithms should be adjusted, and the data sets reevaluated.
- **Adopt Transparent Reporting:** Maintain transparency about how AI tools work, the data they use, and the measures taken to ensure fairness. This openness helps build trust and allows for a more informed dialogue about technology's role in education.

By integrating these strategies, you actively work towards creating a learning environment where AI tools support all students equally, enhancing the educational landscape rather than complicating it.

2.10.2) Advocating For Inclusive AI Design

Inclusive AI design means using AI tools that embrace the full spectrum of human diversity. When AI tools are being realized, they should be fed with data that reflects a wide range of ethnicities, languages, socio-economic backgrounds, and other variables.

As a front-line person in education, your advocacy for inclusive design is powerful. It's about advocating for AI tools that accommodate diversity and are built from the ground up. *This could involve participating in pilot programs for new AI tools, providing feedback to developers, and choosing products from companies that prioritize diversity in their design philosophy.*

2.10.3) Demanding Equitable Access To AI Resources

As referenced earlier, AI's ability to offer personalized learning paths that adapt in real time to the diverse needs of students is the Holy Grail of AI in education. This personalization goes beyond just academic ability. It extends to learning styles, pace, and even cultural contexts.

AI can analyze a student's learning history and adapt the curriculum to focus more on areas where the student struggles by using approaches that foster their visual, auditory, or kinesthetic learning style.

AI can facilitate language translation services and accessibility features for students with special needs. This helps break down barriers that have traditionally hindered these students from fully participating in educational experiences.

Features like speech-to-text for students with dyslexia or real-time captioning for those who are hearing impaired are just a few examples of how AI can be leveraged so each student has access to, and is included in, the same learning opportunities. (**Note**: More on *Special Needs Education* follows).

Primarily due to money, access to technology varies greatly across schools and districts. This leads to gaps in educational equity. As AI tools are increasingly integrated, every student must have the same opportunity to benefit from these advances. It is a critical priority.

Advocating for school policies that provide the necessary hardware and software for all students is paramount. Innovations such as mobile tech labs that visit under-resourced areas can be a short/mid-term, workable solution when optimal funding levels are unavailable.

Also, not all students have access to reliable internet or computing devices in their homes. That being the case, compile a list of community resources, such as public libraries or community centers, where students can freely access AI educational tools.

Likewise, organize after-school programs that provide greater access to existing AI tools so students get a chance to engage with this technology under your guidance.

2.10.4) Cultural Competence in AI Applications

Finally, the AI tools we use must be culturally competent. Meaning, they **should affirm the varied cultural identities of students and avoid perpetuating stereotypes.** An AI system used for teaching history, for example, should provide a multiplicity of perspectives, not just a single narrative. It should recognize and celebrate the contributions of all cultures to our global society.

It is beneficial to review the content delivered by AI tools and make certain it is culturally sensitive and accurate. If an AI tool offers customization options, use these to tailor content that reflects the diverse cultures represented in your classroom. Doing so improves learning and helps students feel respected and connected to their heritage. This, in turn, boosts engagement and academic confidence.

By focusing on these four crucial aspects - mitigating biases, advocating for inclusive design, ensuring equitable access, and enforcing cultural competence - we can steer the integration of AI in education towards a fairer, more inclusive future.

Remember, every tool you introduce in your classroom holds the potential to shape perceptions and influence outcomes academically, socially, and culturally. Choosing and using AI responsibly reflects your commitment to educating equitably and with respect for all the students you serve.

2.11 ONE FINAL OVERARCHING ETHICAL ISSUE: TRANSPARENCY

Introducing AI into our schools and establishing clear and transparent policies around its introduction and use are paramount. Transparency is the cornerstone on which trust arises among all stakeholders - educators, administrators, parents, and students. As such, the question then becomes, "*What does an AI transparent policy look like?*"

It should:

- articulate clearly the purpose of using a particular AI tool,
- clarify the types of AI being employed,
- explain how personal data will be utilized; and,
- explain the measures in place to protect this data.

It should be:

- readily accessible,
- easily understandable; and,
- available in multiple languages to cater to diverse school communities.

To garner support and build trust among those skeptical about AI's role in education, it is powerful to show how AI can positively impact learning. Here are three (3) tangible examples:

Personalized Learning Enhancements: Let parents know that AI can analyze vast amounts of data on students' learning habits and challenges, enabling it to offer tailored educational experiences for their children. Such personalization leads to improved academic outcomes and greater parent and student satisfaction.

Efficiency in Administrative Tasks: Let parents know that AI can significantly reduce the time required to complete administrative tasks and create more opportunities for you to engage with their children. For example, AI-driven systems that automatically grade multiple-choice tests, assess some types of written responses, and provide instant feedback that you can share one-on-one with students.

Enhanced Accessibility: Let parents know that AI tools can assist students with disabilities by offering learning aids such as speech-to-text for those with difficulties in writing, or audiobooks and interactive learning modules for students with reading challenges. Such tools help in fostering access to education and highlight the inclusive potential of AI.

Finally, establishing clear and substantial channels for AI feedback from every stakeholder is essential. Digital surveys, suggestion boxes, or regular focus groups where participants can voice their experiences, concerns, and suggestions are options. This feedback helps gauge the effectiveness of AI implementations and can highlight areas for adaptations and improvements.

By focusing on transparency, collaboration, and continuous improvement, you are not just implementing technology; you are cultivating an informed and engaged school community ready to navigate the future of education together.

2.12 TWO CASE STUDIES OF ETHICAL DILEMMAS AND SOLUTIONS IN AI EDUCATION

To complete our journey into the ethical concerns surrounding AI use, let's examine two (2) real-world scenarios where educators navigated the tricky waters of this topic.

These case studies illuminate potential ethical dilemmas you might face, while offering practical solutions and valuable lessons.

Elementary School Case Study: Balancing Personalized Learning and Privacy

In a bustling elementary school, a well-intentioned initiative to implement an AI-driven personalized learning platform was introduced. This tool was designed to adapt to each student's learning style and pace, promising a revolution in customized education.

However, it was not long before concerns emerged about the amount and type of data being collected. Teachers and parents were worried about the privacy of students' performance data and how securely it was being handled.

The solution involved a collaborative approach between the school's administration and the AI vendor. They revised the data privacy policies to:

- confirm that all data collected was essential for educational purposes and securely encrypted,
- implement stricter access controls; and,
- conduct regular audits to monitor data security.

Finally, Information sessions were held for parents and teachers, explaining the benefits of personalized learning while ensuring transparency about data usage and protection measures.

Lessons Learned

This case highlights three (3) key insights:

1. **Transparency is crucial**: Keeping all stakeholders informed about the AI tools being used and the data they collect helps build trust and eases concerns about privacy.
2. **Collaboration enhances problem-solving**: Working together with AI vendors and involving the community in discussions around AI applications can lead to better, more secure implementations.
3. **Regular audits are essential**: Setting up routine checks on how data is stored and used helps maintain high standards of data privacy and security.

High School Case Study: Addressing AI Bias in College Recommendation Systems

At a diverse urban high school, an AI system was introduced to help counselors provide personalized college recommendations to students. However, it soon became apparent that the AI was not suggesting a broad range of college options equitably to all students. Investigations revealed that the AI's algorithms were inadvertently replicating historical biases present in the data used to inform the neural network.

The school responded by working with AI experts to retrain the system using a more diverse and comprehensive dataset. Further, counselors were trained to review all AI - generated recommendations for any signs of bias before discussing them with students.

This dual approach ensured that students received fair and unbiased college recommendations.

Lessons Learned

From this scenario we learn that:

1. **Vigilance against bias is necessary**: Always be alert to the potential for AI to replicate existing biases and take proactive steps to counteract this.
2. **Human oversight is irreplaceable**: AI is a tool to assist educators, not replace them. Human judgment is crucial, especially in sensitive areas like student counseling.
3. **Continuous improvement is key**: Regularly updating AI systems and training data can help mitigate biases and improve the system's recommendations over time.

2.13 AN ETHICAL DECISION-MAKING EXERCISE

To further engage with these scenarios, here is a short exercise. Imagine you are implementing a new AI tool in your classroom:

1. **List potential ethical issues** you might encounter with this tool.
2. **Propose solutions** to address these issues.
3. **Discuss** your ideas with a colleague or in a professional development session.

This exercise helps you prepare for potential challenges and encourages a deeper understanding of AI's ethical considerations in education.

These case studies are practical examples of how ethical dilemmas in AI education can be navigated. They remind us that while AI offers transformative potential, it also requires careful, thoughtful implementation. By learning from these examples, you can better integrate AI technology into your classroom, and school at large.

Moving forward, please carry these lessons into ongoing discussions about AIs' role in education. This approach benefits all stakeholders and establishes impeccable standards for ethical technology use.

As we transition from ethical dilemmas to anxieties arising from AI use, keep these ethical insights in mind. They are the building blocks for responsibly harnessing AI's capabilities in your classroom.

3

AI - FROM ANXIETY TO CONFIDENCE

Recent studies indicate that the top five (5), AI related anxiety inducers for educators are:

Lack of Understanding or Training

Many teachers feel anxious because they don't fully understand how AI works or how to use it effectively. The fear of *"getting it wrong"* or being outpaced by more tech-savvy peers or students can be intimidating.

"What if I don't know how to troubleshoot it? What if the students know more than I do?"

Data Privacy and Ethics Concerns

Teachers worry about how AI tools collect and use student data. Questions around FERPA compliance, informed consent, and ethical use of student information can create hesitation.

"Is this tool collecting personal student data? Are we staying compliant with school and district policies?"

Fear of Replacing Human Teaching

There's a common concern that AI might diminish the teacher's role or lead to reduced human interaction in the classroom. Teachers often feel the need to prove that real educators can't be replaced.

"Will AI make me less relevant? Will administrators expect it to do my job?"

Increased Workload and Time Investment

Learning a new tool, aligning it with curriculum goals, and troubleshooting technical issues can feel like "*one more thing*" on an already full plate.

"I barely have time for grading—how can I learn this too?"

Equity and Accessibility Worries

Teachers are concerned about whether all students have equal access to AI tools, especially those from under-resourced communities or with special learning needs.

"What if some students can't use this at home? Will it widen the gap?"

Do any of these concerns resonate? Here are a few more:

- natural anxiety about the unknown,
- misconceptions about what AI can or cannot do; and,
- worry about AI making decisions over which there is no control.

Addressing these legitimate concerns head-on means providing accurate information and fostering understanding and trust through dialogue and transparency.

With the goal of exploring, and hopefully dissolving, concerns you may have regarding AI use, we begin by addressing the elephant in the room - adequate funding.

Educators, administrators, school boards, local and state governments must all be involved in the decision-making process when it comes to fully realizing AI equipped schools. Rallying support from all these players takes time, energy, knowledge and focus. Only a collective effort by all the players involved will satisfactorily address this issue. Long term, the efforts are worth it.

For educators who wish to act more quickly - in the here and now, at the local school level – this site may be of great assistance the *Donors Choose* **https://donorschoose.org**

3.1 AN INCLUSIVE DECISION-MAKING PROCESS

It is crucial to involve all stakeholders in the decision-making process. Teachers, administrators, students, and parents need to know the whens, whys, wheres and hows of integrating AI tools into the educational process.

Consider setting up a committee, conducting a focus group, or having a public forum that includes a diverse cross-section of your school community. Prior to the event, think of an AI tool you're interested in using and anticipate the potential concerns each stakeholder may raise, and possible replies. Initial questions may include:

- What key concerns will I cover, unique to each stakeholder?

- How will I demonstrate the benefits gained from the tool?
- How will I address privacy and security concerns?

Such an event helps you prepare for real-world implementation, making the transition smoother and more successful. It can provide valuable insights and foster a sense of ownership. When people feel their voices are heard and concerns addressed, resistance can turn into cooperation and even advocacy.

By addressing anxieties head-on and turning them into informed discussions about benefits and safeguards, you pave the way for AI to become a valuable ally.

3.2 PROFESSIONAL DEVELOPMENT AND TRAINING

To transform apprehension into acceptance, comprehensive development and training are essential. Training programs that speak directly to the benefits, limitations, and ethical considerations arising from AI can demystify it, calm nerves and boost confidence.

Engage with online communities of educators who are using AI in their classrooms; they can be invaluable resources with their advice, compassionate support, and experience sharing. Whether it be a webinar, a forum discussion, or a collaborative virtual workshop, each interaction can significantly shortcut your learning curve and help you make the most of AI technologies. (**NOTE**: see Chapter 5 - *Professional Development and Community Building*).

3.3 SELECTING THE RIGHT TOOLS

Selecting AI tools is akin to curating a gallery of fine art, whereby you select tools that will educate, resonate with, and inspire your students. When selecting your tools for classroom use, it is important to consider several factors that enhance learning while ensuring safety and engagement. Here are six (6) key tips:

3.3.1) Align the Tool with Learning Objectives

Ensure the AI tool supports specific learning outcomes. For example, if the goal is to enhance problem-solving, choose an AI tool that fosters critical thinking or allows students to practice targeted skills, like adaptive math platforms or interactive storytelling systems.

3.3.2) Check for Age Appropriateness

AI tools should be suitable for the students' age group. Look at the content, language, and interface to ensure they are appropriate for your students' developmental level, making the tool engaging but not too complex.

3.3.3) Prioritize Data Privacy and Safety

Review the privacy policies of the AI tool. Ensure that student data is protected, and there is compliance with relevant regulations like FERPA (Family Educational Rights and Privacy Act) or GDPR (General Data Protection Regulation).

3.3.4) Ease of Use and Accessibility

The AI tool should have a user-friendly interface that students and teachers can navigate easily. Additionally, ensure the tool is accessible for students with diverse needs, offering features like text-to-speech, closed captioning, or visual adjustments. Avoid steep learning curves that make integration stressful.

3.3.5) Evaluate Teacher Support and Integration

Check whether the AI tool offers support resources for teachers, such as guides, lesson plans, or training modules. Tools that integrate smoothly with your current technology and curriculum reduce the learning curve and increase classroom effectiveness.

3.3.6) Cost-Effectiveness

Consider the tool's cost relative to the benefits. Expensive doesn't always mean better. Aim for tools that offer significant educational value without breaking the bank.

Incorporating AI to meet your teaching needs should not be a leap into the unknown. With proper research and a thoughtful approach to selection, AI becomes an integral part of your educational toolbox. Fundamentally, AI selection is about:

- making teaching easier and more effective,
- making learning more personalized and effective; and,
- making education more accessible and enjoyable for every student in your classroom.

With these confidence-building ideas in mind, let us now turn our attention to integrating AI at the classroom level.

4

INTEGRATING AI INTO YOUR LESSON PLANS

Three (3) overarching ideas guide the integration of AI into lesson plans. They are:

- confidence in your abilities and AI tools,
- informed preparation that blends curriculum goals and needs with the power and capabilities of AI; and,
- a dash of creativity to customize the learning experience into a memorable classroom experience.

4.1 TIPS FOR OVERCOMING INTEGRATION CHALLENGES

Prior to full-fledged classroom integration, cultivate a comfortable understanding of the capabilities and limitations of your selected tools. Over 150+ are shared throughout the text. Play around with them. Have fun! Learn how extraordinary and impactful this technology can be when utilized effectively.

- follow educational technology blogs,
- explore on-line webinars; or,
- participate in forums where you can exchange ideas and experiences with fellow educators.

Seek out training as required - many providers of AI educational technology offer invaluable training sessions. Contact them.

Explore partnerships with local businesses or tech companies - many organizations are eager to support educational initiatives and can provide the necessary technology or funding to help bridge existing gaps.

Perhaps you will need to advocate for necessary resources. This might mean:

- getting your colleagues on board,
- discussing with Administrators the need for better hardware or software support; or,
- applying for grants and other funding opportunities aimed at technological improvements.

Such advocacy benefits you, your classroom and spearheads broader, institutional change.

Whether the tools are admin or classroom focused, introduce them at a pace that does not foster anxiety. Get comfortable with, not overwhelmed by, your choices. That said, let us consider some practical ways to accomplish this task starting with an administrative focus.

4.2 AI ADMIN TOOLS FOR YOUR TOOLBOX

Is grading taking up much of your weekend? Are lesson preparations eating into your evenings? Once you pinpoint these time eaters, seek AI tools designed to alleviate these specific burdens. There are numerous AI-powered tools that can:

- evaluate student assignments,
- grade quizzes,
- provide detailed feedback and insights into student understanding,
- help you generate and customize lesson plans based on your goals and needs.
- monitor student progress in real time, alerting you to whom might need extra help or which concepts need reiteration,
- manage ordering supplies and materials,
- scheduling parent-teacher conferences,
- automate the sending of report cards; and,
- track student attendance, then analyze patterns of absenteeism and provide insights that aid in improving student participation.

Once mastered, the following tools reduce the administrative load, allowing you to focus more on teaching and less on paperwork.

4.2.1) Google Classroom: **https://edu.google.com/workspace-for-education/products/classroom/**

Best for: Grading, communication, assignments

Google Classroom lets teachers create, distribute, and grade assignments all in one place. It also integrates with Google

Calendar for scheduling and Docs/Sheets for reports. It's free and simple to use, especially for schools using Google Workspace.

4.2.2) ClassDojo: **http://www.classdojo.com**

Best for: Parent communication, attendance, behavior tracking

ClassDojo makes it easy to connect with parents, send updates, track attendance, and even document student behavior. It creates a positive communication loop between school and home—ideal for scheduling check-ins or sharing student progress.

4.2.3) PowerSchool: **https://powerschool.com**

Best for: Report cards, attendance, student data management

PowerSchool is a full-featured student information system (SIS). Teachers can manage grades, attendance, report cards, and access student records. It's widely used by school districts and supports custom reporting tools for administrators.

4.2.4) Planbook: **https://planbook.com**

Best for: Lesson planning, scheduling, curriculum tracking

Planbook helps teachers organize their lesson plans and schedules. You can align lessons to standards, share plans with other teachers, and stay on track. It also allows flexible adjustments for unexpected schedule changes.

4.2.5) Khanmingo Your TA: **https://khanmingo.ai**

Best for: Offers over 25 educator tools in a tidy, easy-to-navigate interface

Bottom line, front-end, admin AI integration that focuses on:

- time management,
- efficient tool selection,

- easy to use applications; and,
- will streamline your work, leaving greater space for student interactions and teaching.

4.3 TARGETING CURRICULUM WEAK SPOTS WITH AI

Pinpointing the perfect moments for AI integration and creating flexible learning modules that adapt to both your curriculum needs and time constraints require a foundational understanding.

Start by identifying those points in your curriculum where students typically struggle or where you spend a lot of time on grading and personalized feedback. These are prime spots for AI integration.

If you notice that students often struggle with complex algebraic concepts, integrate an AI tool that offers personalized quizzes and instant feedback.

Similarly, AI-driven content generators can help Language Arts students better understand grammar or overcome writer's block by suggesting ideas and structures for their essays.

By mapping out where AI tools can be most effective, you ensure they add value without disrupting the flow of your existing curriculum. Here are five (5) tips to assist you.

4.3.1) Use AI to Analyze Student Data for Patterns

AI tools like learning management systems or platforms like *Edpuzzle* (**https://edpuzzle.com/**) and *Quizizz* (**https://quizizz.com/join/dashboard**) can help track student performance. Use this data to pinpoint which topics consistently challenge your class — whether it's fractions, grammar rules, or scientific concepts.

Once you know the weak spots, you can focus your instruction more effectively.

4.3.2) Leverage AI for Differentiated Instruction

Every student learns differently. AI can help you generate tailored content — from simpler explanations for struggling students to enhanced content for advanced learners.

Tools like *ChatGPT* and *MagicSchool.ai* (**https://www.magicschool.ai/**) are great for generating age-appropriate reading passages, vocabulary work or tiered practice questions.

For example (as a teaser) pause and enter the following prompt into *ChatGPT* and see what you think of the results. (**Note**: Please know that Chapter 7 focuses on effective AI activity creation).

"for Grade 3, Grade 6 and Grade 8, provide one printable worksheet activity for each grade level that teaches an age-appropriate math question about multiplication AND provide a separate answer key for each activity"

4.3.3) Create Smart, Adaptive Practice

AI-based learning platforms like *Dreambox* (**https://www.dreambox.com/**) and *Sora* (**https://soraapp.com/welcome**) offer real-time adjustments based on student responses. These tools reinforce concepts and give students a chance to practice where they need the most help, without always relying on your direct intervention.

4.3.4) Generate Quick, Targeted Review Materials

Once you have identified weak spots, i.e., past tense verbs or converting units, let AI help you quickly create focused review sheets, games, or visual aids. You do not have to continually re-invent the wheel. (**Note**: More about this in Part Two).

4.3.5) Encourage Students to Use AI as a Learning Partner

Teach students how to responsibly use AI to reinforce their own understanding - like asking clarifying questions, getting feedback on their writing, or practicing vocabulary. This builds their independence and tech fluency while helping them revisit tough concepts. Share how to ask AI a good learning question, like: "*Can you explain how to simplify fractions in a simple way?*"

4.4 SATISFYING EDUCATIONAL GOALS

When gearing up to integrate, review your existing lesson plans line-by-line and identify the specific learning goals. Ask yourself: *On completion of the topic at hand, what will my students have learned? What type of AI activities will support achieving this goal?*

For example, if you are teaching middle school biology with the goal of understanding the life cycle of a frog, GPT Chat prompt:

"for grade 6 level, create a printable activity worksheet about the life cycle of a frog in 10 steps and generate a 10 question, multiple choice quiz about it AND a separate answer key"

Or

"for grade 6 level, create a printable activity worksheet about the life cycle of a frog with 10 mixed up sentences describing the cycle that students must order correctly AND a separate answer key"

See what AI produces. Do not like it? Alter the prompt and try again until you are satisfied.

4.5 DESIGNING INTERACTIVE ACTIVITIES

With goals defined, begin designing engaging, interactive activities that make learning with AI so compelling. Imagine you are using *ChatGPT* to simulate historical events for a social studies class.

For example, after teaching the history surrounding 1960's, America, JFK, his policies, etc... you engage students with the results of the following prompt:

Prompt: *for grade 9, in 300 words, if President John Kenedy had not been killed in 1963, what could have been 3 possible consequences for the country?*

This can result in some terrific learning moments and speculations. (**Note**: More about this in Part Two).

4.6 MODULAR AI LESSONS

In need of some flexibility? Modular AI lessons may help. They are like handy travel kits - packed with essentials, ready to go, and easy to fit into your luggage.

Designing AI-integrated lessons in a modular format allows you to easily shift them throughout your teaching schedule wherever they fit best. Each module can be self-contained, focused on a specific topic and complete with AI-enhanced activities and assessments.

For example, you could have a 3-part module on a particular time frame, for example, 1900 to 1950 American History. Part One requires students to research economic conditions. Part Two looks at impactful inventions during the era, and Part Three explores social relations and immigration.

Another module could focus on scientific experiments with AI simulations that allow students to observe chemical reactions in a virtual lab.

The beauty of modular design is that you can use these lessons as stand-alone enhancements or linked together for a more comprehensive AI-driven unit.

Further, if a particular module is not fitting well with your current dynamics of the class or if time constraints tighten, you can easily remove it without disrupting the overall curriculum flow. This flexibility is invaluable in today's educational environment, where adaptability can often be as crucial as the content itself.

As an example, enter the following prompt into ChatGPT, or your preferred AI source, and see what you think.

prompt: *using the topic of photosynthesis for grade 5 level, provide a printable, AI-integrated lesson in a modular format that can be easily shifted throughout a teaching schedule wherever it best fits. Each module can be self-contained, focus on a specific sub-topic complete with AI-enhanced activities AND separate answer keys*

Here Is The Result

Here is a Grade 5 printable, AI-integrated modular lesson on Photosynthesis. It's structured for flexible use, with self-contained modules that include a brief intro, AI-enhanced student activities, and separate answer keys.

LESSON TITLE: Understanding Photosynthesis

Grade Level: 5

Format: Modular | Printable | AI-Integrated

MODULE 1: What is Photosynthesis?

Objective: Define photosynthesis and understand its basic purpose.

Mini-Reading: Photosynthesis is how green plants make their own food. Using sunlight, carbon dioxide from the air, and water from the ground, plants create glucose (a type of sugar) and release oxygen into the air.

Activity – AI Prompt (for ChatGPT or similar): *"Explain photosynthesis to a 7-year-old using a story about a plant superhero named Captain Greenleaf."*

Students write or type their version of the AI-generated story and compare it to their own.

Follow-Up Questions (Circle the correct answer):

1. What do plants need for photosynthesis?
2. a) Sugar, Oxygen, Water
3. b) Sunlight, Water, Carbon Dioxide
4. c) Soil, Air, Heat
5. What is the main product of photosynthesis?
6. a) Protein
7. b) Glucose
8. c) Carbon

MODULE 2: Where Does Photosynthesis Happen?

Objective: Identify plant parts involved in photosynthesis.

Mini-Reading: Photosynthesis mostly happens in the leaves. Inside leaves are tiny parts called chloroplasts. These have a green pigment called chlorophyll, which captures sunlight.

AI-Enhanced Activity Prompt: *Draw a simple diagram of a leaf and ask AI to label the parts involved in photosynthesis.*

Students sketch and label: leaf, chloroplasts, sunlight, water, CO_2, glucose, oxygen

Short Answer:

1. What part of the plant collects sunlight?
2. What does chlorophyll do?

MODULE 3: The Photosynthesis Formula

Objective: Learn and break down the basic chemical formula of photosynthesis.

Mini-Reading: The formula looks like this:

$CO_2 + H_2O + \text{sunlight} \rightarrow C_6H_{12}O_6 + O_2$

Carbon dioxide + water + sunlight → glucose + oxygen

AI Prompt for Practice:

Turn the formula for photosynthesis into a poem or rap explaining each part.

Students create short rhymes or AI helps them write it.

Fill-in-the-Blank:

1. Plants take in ____________from the air.
2. They get ____________from the soil.
3. They use ____________from the sun.
4. They make ____________(sugar) and release ____________.

MODULE 4: Why is Photosynthesis Important?

Objective: Explain the impact of photosynthesis on life and the environment.

Mini-Reading: Without photosynthesis, humans and animals wouldn't have oxygen to breathe. It also starts the food chain by making plant food.

AI Prompt for Discussion: *Pretend you're a scientist on Mars explaining why Earth is special because of photosynthesis.*

Students write a paragraph using AI assistance.

Matching Activity:

Match the benefit to the effect:

- Gives oxygen → ______
- Starts food chains → ______
- Removes carbon dioxide → ______

Options: a) Less CO_2 in the air, b) Animals breathe, c) Plants feed animals

Separate Answer Keys

Module 1:

1. b
2. b

Module 2:

1. Leaf
2. Captures sunlight for photosynthesis

Module 3:

1. Carbon dioxide
2. Water
3. Sunlight
4. Glucose, Oxygen

Module 4:

- Gives oxygen → b
- Starts food chains → c
- Removes carbon dioxide → a

Would you like this in a printable PDF format with spaces for name/date and illustrations included?

4.7 TWO TYPES OF FLEXIBLE AI INTEGRATION

Flexibility should be a core consideration of AI integration. Not every class or school year is the same, and your approach to using AI tools should adapt to changing circumstances. Think of it as having a set of adjustable wrenches in your toolbox, each capable of tightening or loosening to meet the specific needs of a job.

One effective model is **on-demand AI integration**. Here, AI tools are available for students to use as needed. For example, an AI-powered writing assistant can help students structure their arguments, while others might choose to consult it for final grammar checks.

Another model is **embedded AI integration**. Here, AI is linked with the learning platforms students use every day. Such integration places AI tools as a natural part of the learning process, as opposed to an add-on that may feel out of place.

Both models emphasize AI as a support tool that makes it easier for you to adopt AI and helps your students see it as a normal part of their learning environment.

By taking the time to map AI integrations thoughtfully - designing flexible/ adaptable modules - AI becomes a seamless and stress-free component of your educational toolbox. These strategies allow you to harness AI benefits in a manner that respects your time constraints and teaching goals.

4.8 AI INTEGRATION STRATEGIES FOR TIGHT BUDGETS

Integrating AI into your classroom needs not be a budget buster. Commonly, maximizing what you already have, rather than acquiring new resources, can be the key.

Explore existing resources and see if there are any underutilized digital tools already available. Many educational platforms offer basic AI functionalities like data analysis or personalized learning insights. You might discover that you are already equipped to start incorporating AI into your lessons at no extra cost.

Additionally, there are many effective AI tools available online for free. Earlier, I mentioned **Google Classroom.** It uses AI to help teachers organize assignments, boost collaboration, and manage classes efficiently, all without an additional fee.

Khan Academy (**https://www.khanacademy.org/**) offers students personalized learning paths across a range of subjects - from mathematics to the arts and humanities. While there, be sure to access their teaching resources, too. Their offerings are terrific, and everything is *funded through donations.*

Duolingo (**https://en.duolingo.com/**) is an AI-driven, language learning platform that adapts to each user's learning style and pace, and *it's free for basic use.*

For those interested in more content-specific tools, *PhET Interactive Simulations* (**https://phet.colorado.edu/**) provides *free science and math simulations.* It adapts scenarios of complex concepts through virtual experiments based on student interactions.

Integrate AI incrementally. Start small by introducing one or two AI tools that address specific needs or challenges in your classroom. This could be as simple as using an AI-powered quiz tool to assess student understanding, or a chatbot to help answer students' frequent questions after school hours.

Gradually, as you and your students become more comfortable with these tools, expand your AI toolbox. This makes the process manageable and allows you time to evaluate the effectiveness and impact of each tool.

By leveraging existing resources, cultivating community support, utilizing free or low-cost AI tools, and adopting an incremental integration approach, you can effectively and affordably bring AI into your classrooms.

These strategies ensure that financial constraints do not hinder your ability to provide a modern, technology-enhanced education that prepares students for the future. (**Note**: Chapter Six has an expanded list of resources based on grade levels and topics.)

4.9 AI AND THE LOSS OF HUMAN TOUCH

In the rush to integrate AI, we must never forget that **we are the heartbeat of teaching**. There is a genuine concern among educators that the essential human connection which nurtures students' growth could be overlooked.

4.9.1) Balancing Technology with Humanity

AI is a tool to support, not supplant, the interactions that make your classroom a vibrant learning environment. While an AI program might help a student with grammar, it is your discussions that arise from their writing that can challenge and inspire deeper creativity and expression.

AI can analyze a student's performance and pinpoint areas of difficulty, but it's your encouragement and personal guidance that can motivate them to try harder and become better. **We must preserve the personal touch that is crucial in nurturing a supportive and engaging learning environment by establishing clear boundaries and guidelines for healthy AI use**.

AI should encourage collaboration and interaction among students, not provide yet more screen time. They get enough of that on their phones and tablets. AI facilitated group projects and interactive simulations that require teamwork, and class discussions keep the human element alive and thriving in your classroom.

4.9.2) Cultivating Emotional Intelligence

Programs and strategies that focus on *Emotional Intelligence* (EI) are vital in ensuring students learn empathy, self-awareness, and effective communication alongside their academic subjects. Used properly, AI can support these goals.

AI-driven analytics can help you understand students' emotional states by analyzing patterns in their engagement and feedback. This insight allows you to tailor your interactions more effectively, as you check-in with a student who seems to be struggling, or guide role-playing activities whereby students navigate through social scenarios and practice empathy and problem-solving.

4.9.3) Another strategy is to integrate *Social-Emotional Learning* (SEL) **https://beyou.edu.au/fact-sheets/social-and-emotional-learning/five-areas-of-social-and-emotional-learning** programs that explicitly teach students about emotions, how to manage them, and how to build positive relationships. Explore another SEL resource by entering *The Zones of Regulation* into your search engine and see what you think.

When integrating AI into your teaching practices, aim to illuminate the human aspects that make education a transformative experience. Always remember that the goal is to prepare students, not just for the marketplace, but for the complexities of life itself.

4.10 SIMPLIFYING AI CONCEPTS FOR YOUNG LEARNERS

Explaining complex AI concepts in ways that resonate with young learners is best achieved by starting with things they already know - things like Siri or Alexa. Let's explore some creative, age-appropriate methods to help young students grasp the basics of AI, making it both understandable and engaging.

4.10.1) Age-Appropriate Explanations

Sharing examples and metaphors from their everyday experiences are an excellent path forward towards explaining AI.

A great example is relating AI to video games. Start with a discussion about their favorite games. Then, when the time is right, ask them if they know how video games work. This opens the door for explaining that their AI opponent is making decisions based on their own moves – just like people playing chess.

Or, metaphorically, you can compare AI to their favorite cartoon character; they talk, play and make decisions. Explain that just like their cartoon friend, AI can listen, learn, and even think up ways to solve problems within the cyber world.

4.10.2) Interactive AI Activities

To truly bring AI concepts to life for younger students, interactive activities are incredibly effective. Introducing them to an AI-powered, age-appropriate app that adapts to their learning level and allows them to experience personalized learning. With colorful interfaces and game-like elements to teach basic math or reading skills, this hands-on interaction shows them what AI can do and is.

4.10.3) Visual Aids and Resources

Visual aids are invaluable in teaching complex concepts to young learners. Illustrate how AI works in a simple and engaging way with charts, videos, and infographics. A colorful infographic can show how a voice assistant like Siri or Alexa processes questions to find answers.

Videos, particularly animated ones, can visually explain what code is and how it tells computers what to do, which is fundamental to understanding AI.

4.10.4) Consider using interactive online platforms where children create their own AI models via simple, drag-and-drop interfaces. Platforms like *Scratchjr* (**https://www.scratchjr.org/**) designed for younger audiences, allow students to build and program their own interactive stories and games with AI elements.

All these resources both clarify AI concepts and empower children to create with technology, not simply consume it.

Another fantastic visual resource for introducing topics is **www.YouTube.com.** For example, you want to introduce the topic: *life* cycle of a butterfly. **Access the site and prompt the following:** *for grade 3 level, provide a video about the life cycle of a butterfly.*

This results in multiple options, many of which are only 3-5 minutes, lending themselves to easy repeat viewings.

I encourage you to curate the options and **capture the link of the one you most like. Share it with students so they can watch it as much as they need.** This is particularly helpful for math and science topics. In Part Two, there are several sample lesson plans that integrate this resource.

Further, for elementary students, there are a lot of *sing and learn* videos that are terrific. Check this one out by entering this prompt in the **YouTube** search bar: *Place Value Song For Kids | Ones, Tens, & Hundreds | 1st - 3rd Grade/ Math Songs by Numberock*

(**Note**: You will have to click *"skip this"* to avoid the ad that precedes free access).

4.11 ENGAGING RELUCTANT LEARNERS WITH AI

As mentioned earlier, AI's ability to create personalized learning experiences is not a hopeful vision, it is already changing how we teach.

With systems designed to analyze large sets of learning data, such as:

- test results,
- time required to answer a question; or,
- the types of questions with which students struggle.

Personalized learning now extends well beyond these simple, academic metrics.

These systems can also pick up on what themes or activities spark student engagement. A student who might be reluctant to engage with traditional methods of learning history might show more interest when the content is delivered through an interactive AI-powered virtual reality tour of ancient Rome.

The AI system adapts, noting this preference, and integrates more visually immersive content into their learning path, gradually increasing their engagement and interest in the subject. Such personalization extends to pacing.

Every student learns at their own pace, and traditional classroom settings can often make reluctant learners feel rushed or, conversely, held back. AI's adaptability allows for a learning tempo that matches perfectly to each student's learning speed, offering more challenging material when they're ready or extra time and resources when they're struggling.

This can be particularly empowering for students who might have previously felt out of step with the standard classroom pace as they engage more fully and confidently with the material.

4.12 GAMIFICATION TECHNIQUES

Turning learning into a game is a time-tested method to boost student engagement. With AI is thrown into the mix, the possibilities for gamification become even more exciting and effective. Let's explore how AI leverages gamification techniques to capture the interest of those students who might otherwise be disengaged.

AI can be used to create dynamic learning environments that adjust in real-time to the actions and decisions of the student. When learning a new language, an AI system can introduce competitive elements such as points, levels, and badges based on the student's progress in real-time quizzes and pronunciation exercises.

These systems can also introduce challenges that are directly tailored to the student's current level, pushing them just enough to keep the learning process challenging and engaging without leading to frustration.

Other AI-driven simulations and role-playing games can immerse students in scenarios that require critical thinking and problem-solving skills.

This type of interactive learning makes the subject matter more relatable and interesting, allowing students to see the real-world applications of what they are learning. This makes them realize the actual value of their educational activities.

AI gamification can also enhance collaboration. Multiplayer educational games, moderated and adjusted by AI, can encourage students to work together to solve problems or complete projects. By combining social interaction with learning, the entire process becomes a shared adventure rather than a solitary task.

The integration of AI-driven gamification into learning strategies presents a significant opportunity to transform traditional education into a more engaging, interactive, and enjoyable experience.

For reluctant learners who might find traditional educational methods daunting or dull, these AI-enhanced games and simulations can be the key to unlocking their enthusiasm and potential. AI has the power to turn reluctance into curiosity, and curiosity into a passion for learning by making learning fun and tailored to their interests and needs.

Here are three (3) popular gamification sites.

1. *Kahoot!:* **https://kahoot.com**

Kahoot! turns classroom review and learning into a fast-paced game show. Teachers can create or use existing multiple-choice quizzes called "*kahoots*," and students can join via a game PIN using any device. The leaderboard and music add excitement, making it a favorite from elementary to high school.

2. *Prodigy:* **https://www.prodigygame.com**

Prodigy is a role-playing game (RPG)-style math platform that covers content from Grades 1–8. Students explore virtual worlds, cast math-powered spells, and complete curriculum-aligned challenges. It's free to use, with optional paid features, and adapts to each student's level.

3. *Classcraft*: **https://www.classcraft.com**

Classcraft gamifies the classroom itself. Students create characters and earn points for positive behavior, collaboration, and academic achievements. Teachers can set quests, battles, and team missions.

It's especially effective in upper elementary through high school for improving engagement and classroom management.

4.13 AI FOR SPECIAL NEEDS EDUCATION

In the realm of special education, AI's potential for aligning learning experiences with individual needs is transformative by customizing learning tools. This ranges from adjusting the content and pace of learning to modifying the delivery methods to suit different sensory needs.

AI-powered platforms can dynamically adjust the complexity of text for students with dyslexia, providing simpler synonyms and sentence structures, or enhance contrast and font size for those with visual impairments.

For non-verbal students or those with speech and language difficulties, AI offers invaluable tools to bridge these gaps using symbols, text, and speech generation.

Some AI speech recognition software can be trained to understand and interpret the unique speech patterns of students with articulation difficulties, providing them with more independence in expressing themselves.

Additionally, chatbots can serve as practice companions for students to improve their verbal and social skills in a low-pressure environment.

Creating an inclusive classroom goes beyond just having the right tools; it involves designing learning experiences that integrate these tools seamlessly into everyday activities, ensuring all students benefit from them.

One effective strategy is using AI to create collaborative projects where students with special needs and their classmates work together, using technology as a bridge.

For instance, an AI-driven project could involve students building a story together where AI suggests plot ideas, grammar corrections, and even illustrations, based on each student's input and interactions.

Another strategy is the use of AI-driven analytics to monitor classroom dynamics and learning patterns, helping teachers identify when students might feel excluded or need additional support.

This data can guide teachers in structuring group activities in ways that promote more balanced participation to ensure that each student's contributions are valued.

As with any educational tool, the efficacy of AI in special needs education heavily relies on the teachers who implement it. Professional development and training are crucial in equipping educators with the knowledge and skills they need to use AI effectively and ethically.

Special needs education teachers can benefit from training programs that focus on the operation and integration of the tools with teaching methods that respect and advance the rights and potential of special needs students.

Adopting AI in special needs education is about harnessing technology and creating a learning environment that adapts to each unique student - not the other way around.

As AI continues to evolve, its role in special needs education aims to make education truly inclusive, engaging, and empowering for all students, regardless of their personal challenges.

Here are several resources to consider:

1. *Common Sense:* **https://www.commonsense.org/education/lists/apps-and-websites-for-special-education-programs**

2. *Microsoft for Special Needs:* **https://www.microsoft.com/en-us/accessibility**

This offers a range of accessibility features and products designed to support individuals with special needs. This includes built-in tools in Windows, like **Narrator, Magnifier, and Live Captions**, as well as adaptive accessories and support programs.

Microsoft also provides tools within *Microsoft 365*, such as **Immersive Reader, Accessibility Checker, and Microsoft Translator**, to enhance productivity and accessibility.

3. *Proloquo2Go:* **https://www.assistiveware.com/products/proloquo2go**

Best for: Non-verbal students or those with speech challenges (e.g., autism, cerebral palsy)

- Augmentative and Alternative Communication (AAC) app
- Uses symbols and text-to-speech to help users communicate
- Customizable vocabulary for different developmental levels

4. Choiceworks: **https://www.beevisual.com/**

Best for: Visual scheduling, emotional regulation, and routines

- Helps students understand daily tasks using picture schedules
- Teaches how to wait, take turns, and manage feelings
- Ideal for students with autism, ADHD, and executive functioning challenges

5. ModMath: **https://www.modmath.com/**

Best for: Students with dysgraphia or difficulty with handwriting

- Free app that lets students do math without writing
- Virtual graph paper for lining up equations
- Supports addition, subtraction, multiplication, division, and fractions

6. Bookshare: **https://www.bookshare.org/**

Best for: Students with reading disabilities like dyslexia or visual impairments

- Provides free access to 1M+ accessible books for eligible U.S. students
- Offers audio, large print, Braille, and customizable font formats
- Syncs with reading apps like *Voice Dream* and *Dolphin EasyReader*

7. *Understood.org:* **https://www.understood.org/**

A comprehensive resource for supporting students with learning and thinking differences such as ADHD, dyslexia, and sensory processing issues. The site provides personalized learning strategies, classroom tips, IEP support, and resources for both educators and parents.

8. *Teaching Students with Special Needs (Reading Rockets):*

https://www.readingrockets.org/teaching/special_education

Reading Rockets offers expert articles, classroom strategies, and practical advice for teaching children with a wide range of learning disabilities. It focuses on literacy but includes tools and insights that apply across subjects and grade levels.

9. *National Center on Accessible Educational Materials (AEM Center):* **https://aem.cast.org/**

This site helps educators provide materials in formats accessible to all learners, including those with physical, sensory, or cognitive disabilities and offers guidance on digital tools, assistive technologies, and inclusive curriculum planning.

Hello, fine reader and educator, a brief pause and request.

"The role of the teacher is to create the conditions for invention rather than provide ready-made knowledge."

— *SEYMOUR PAPERT, A PIONEER IN ARTIFICIAL INTELLIGENCE AND EDUCATION*

May you be finding Power UP Your Classroom With AI accessible, engaging and useful. I had four (4) goals in mind while creating this book. They were:

- Share an understanding of the origins and ethics of AI,
- Provide the knowledge and skills necessary to transform AI anxieties into confidence,
- Teach how to create effective, focused AI prompts to need your needs and goals; and,
- Identify tools to help streamline admin duties & lesson planning.

Towards realizing these goals, I need your help.

People often choose books based on the experiences and insights of others. So, I'm asking you to please consider leaving a review. It costs nothing and takes only a minute, but your words could transform another teacher's STEM journey. Your review could:

- Help one more teacher find confidence to bring STEM to life for their students,
- Help one more student discover new ways to think, learn and grow; and,

- Help one more classroom become a pathway to success in our STEM/ tech-driven world.

If you'd like to help, simply scan the QR code below to leave your thoughts:

Thank you for your support and for helping make education stronger, one classroom at a time.

With gratitude, pjs

PART II

5

PROFESSIONAL DEVELOPMENT AND COMMUNITY BUILDING

This chapter delves into some of the top professional development programs specializing in AI education. It shares the information you need to select the best fit for your educational and professional growth, as well as ideas for building a community of educators.

5.1 TOP PROFESSIONAL DEVELOPMENT PROGRAMS IN AI EDUCATION

Here are four (4), top ranked programs for your consideration.

AI Teaching Guide by Stanford University: This program offers online and in-person workshops focusing on how AI can be applied in various educational settings. From understanding machine learning to implementing AI tools for personalized learning, the program covers a range of AI applications in education.

https://tlhub.stanford.edu/docs/quick-start-guide-to-ai-and-teaching/

Microsoft's AI Education School: Partnering with schools worldwide, Microsoft offers this program to help educators develop practical AI skills. It includes training on AI use for administrative tasks, enhancing student engagement, and improving learning outcomes through data analysis.

(**Note***:* Enter *Microsoft's AI Education School* into your search bar to access the site; the exact link is way too long)

Google's Generative AI for Educators: This program offers a range of AI tools tailored for educational purposes, helping teachers learn how to optimize utilization and include hands-on activities where you can experiment with AI technologies.

https://grow.google/ai-for-educators/

IBM AI Education Series for K-12 Educators: This series provides comprehensive training on integrating AI into various subjects. It teaches the technical skills needed to use AI and how to creatively incorporate AI projects to enhance student learning and engagement.

https://mediacenter.ibm.com/media/IBM+AI+Education/1_frksa038

5.2 CRITERIA FOR SELECTION

How do you choose the program that best serves your needs? Here are several criteria to assist you in the selection process.

- Look for programs that align with your specific educational goals. Whether it's learning about AI tools, understanding AI applications in curriculum development, or grasping the ethical aspects of AI, the program should match your educational needs and priorities.

- Ensure that the professionals conducting the program have a deep understanding of applying AI in educational settings.
- The ideal program should offer practical, hands-on training that you can readily implement in your classroom. While theory is essential, the real value comes from being able to apply what you learn directly to your teaching.
- Post-program support, such as access to resources, ongoing training, and a community of fellow educators, significantly increase a professional development program's value.

5.3 ADDITIONAL DEVELOPMENT RESOURCES

A vast world of online AI resources is rapidly arising. Here are a few.

AI4ALL offers a range of courses covering all aspects of AI.

https://ai-4-all.org/resources-all/

EdX provides courses developed by top universities on AI applications and ethics; within the site, search for "AI" and you will find 1200+ AI related courses.

https://www.edx.org/search?q=AI

ISTE – Artificial Intelligence in Education: The International Society for Technology in Education (ISTE) offers free guides, professional development resources, webinars, and lesson plans focused on AI in K-12 education. It also provides frameworks to teach students about ethical AI use.

https://www.iste.org/areas-of-focus/AI

Teachable Machine by Google is great for teachers who want to explore AI hands-on with students. This free tool allows users to create machine learning models in minutes using images, sounds, or poses - no coding required. It also includes educator guides and tutorials.

https://teachablemachine.withgoogle.com/

AI4K12 was created for K–12 educators and offers AI teaching frameworks, lesson ideas, and age-appropriate learning progressions across five core AI concepts.

https://ai4k12.org/

IBM SkillsBuild for Educators offers free digital courses, certifications, and AI teaching modules. Teachers can access career-aligned lessons on AI ethics, machine learning, and real-world AI applications with classroom resources and student activities.

https://skillsbuild.org/educators

Edutopia – Artificial Intelligence in the Classroom publishes articles, case studies, and teaching strategies for integrating AI tools in learning. Ideal for staying current on best practices and practical tips from other educators.

https://www.edutopia.org/topic/artificial-intelligence

AI For Education: This site has a wealth of material and offers a free, 2-hour introductory course.

https://www.aiforeducation.io/ai-course

5.4 SOCIAL MEDIA COMMUNITIES FOR EDUCATORS

For real-time interaction and support from the AI educational community, here are four (4), active online forums and social media groups for your consideration.

AI for Education (LinkedIn Group): This professional group hosts discussions, shares resources, and networks educators interested in AI from around the globe. It's a great place to find collaboration opportunities and get advice on AI integration in schools.

EdTech AI (Reddit): This subreddit is a forum where educators discuss the latest AI tools in EdTech, share experiences, and provide support for implementing AI in educational settings.

AI4K12 (Facebook Group): This group is dedicated to K-12 educators looking to integrate AI into their curricula. Members share lesson plans, AI tool reviews, and strategies for teaching AI concepts to young learners.

Teach with AI (Forum): An independent online forum where educators post queries, share classroom experiences, and discuss the ethical implications of AI in education. The community is supportive and quick to offer practical solutions and ideas.

5.5 BUILDING A COMMUNITY OF AI-ENHANCED TEACHING

Building a community among passionate educators creates the space for sharing knowledge, resources, and innovative teaching methodologies.

By connecting with fellow educators interested in AI, you are essentially joining a think tank that thrives on innovation and mutual support. In such a community, members can:

- share their triumphs and challenges,
- learn from each other's experiences; and,
- collaboratively develop solutions that might not have been apparent in isolation.

Community engagement makes AI more accessible and less intimidating for educators who might feel overwhelmed with the technical aspects.

By pooling resources and knowledge, the community can navigate the complexities of AI more effectively and confidently, leading to enriched learning environments that prepare you for this new world of learning.

If you are so enthused, here are ten (10) steps towards building a vibrant community.

A) Define a Clear Purpose: *"To support educators in understanding, implementing, and sharing practices around AI-assisted learning tools in the classroom."*

B) Identify Your Audience: Determine who you want to include:

- K–12 teachers
- EdTech coordinators
- Administrators
- Pre-service educators
- AI researchers in education

C) Choose a Platform: Select a space to meet and communicate:

- **Online:** Slack, Discord, Facebook Groups, Google Groups, or a private forum
- **Local:** Monthly meetups at schools, libraries, or educational centers

D) Create Core Content and Resources: Start with useful resources:

- AI tool guides for teachers
- Sample lesson plans using AI
- Case studies of successful AI classroom integration
- Ethics discussion templates

E) Invite Founding Members: Reach out to 5–10 educators who are already exploring AI or are interested. Their early feedback and energy help shape the tone and focus of the group.

F) Establish Regular Events: Start small

- Weekly tip-sharing threads
- Monthly online roundtables or guest talks
- "Tool of the Month" demo and discussion

G) Encourage Sharing and Collaboration: Make it easy for members to:

- Ask questions
- Share lesson ideas or student outcomes
- Post AI-related news or research
- Collaborate on projects or pilots

H) Promote Across Channels: Grow the community through:

- School district newsletters
- Social media posts
- Educational conferences
- Partnerships with EdTech companies or teacher colleges

I) Gather Feedback and Evolve: Regularly ask members what they need

- New tools?
- More classroom examples?
- Training sessions?
- Adapt and improve accordingly.

J) Celebrate and Showcase Success: Highlight real-world classroom success stories using AI. Feature educators, student work, or school programs to inspire and build credibility.

As the community grows, consider appointing roles or creating sub-groups focusing on specific areas of AI education, such as ethics, tools, or curriculum integration.

Finally, establish goals and milestones for the community. These could range from short-term objectives like completing a group training session on a new AI tool, to long-term goals such as co-authoring a paper on AI in education. Shared goals such as these provide a clear direction and foster a healthy sense of accomplishment.

Bearing all these issues in mind, let's now turn attention to sharing resources by grade level and topic, using AI to create focused activities and looking at sample lesson plans.

6

RESOURCES BY GRADE LEVELS AND SUBJECTS

For your consideration …

6.1 MATH

6.1.1) Elementary Level

1. *Prodigy Math Game*: **https://www.prodigygame.com/main-en/teachers**

This is a fantasy-based, curriculum-aligned RPG (role playing game) covering grades 1–8. Adaptive learning, quests, and parent/teacher dashboards. Used by millions worldwide. Free basic version with optional premium.

2. *Khan Academy Kids*: **https://www.khanacademy.org/kids**

This is a free, nonprofit app for ages 2–8 offering thousands of lessons in math (plus language), video tutorials, and interactive practice. Highly trusted and 100% accessible.

4. *DragonBox Numbers*: **https://dragonbox.com/products/numbers**

Developed for ages 4–8, this visually rich app promotes number sense through playful puzzles and activities.

5. *Photomath*: **https://photomath.com**

This is a camera-based math solver for elementary to higher-level math. Students share a photo of their problem and get step-by-step explanations. It is an excellent support tool alongside learning. Free version, plus premium features.

6.1.2) Middle School Level

1. *Khan Academy*: **https://www.khanacademy.org/**

A completely free, nonprofit platform offering lessons and practice in grades 6–8 math - covering arithmetic through introductory algebra and geometry. It includes engaging videos, mastery challenges, and progress tracking tools for students and teachers.

2. *Photomath*: **https://photomath.com/**

See above

3. *DragonBox Algebra*: **https://dragonbox.com/**

This is a game-based algebra introduction where kids transform puzzles into algebraic thinking. Highly praised for making abstract concepts intuitive and engaging via drag-and-drop gameplay.

4. *GeoGebra*: **https://geogebra.org/**

This is a powerful interactive tool for exploring geometry, algebra, and basic calculus. Great for visual learners - students can manipulate shapes and functions dynamically and access thousands of pre-made applets (small apps).

5. *QANDA*: **https://qanda.ai/en**

An OCR-based (Optical Character Recognition) AI tutoring app that lets students scan math problems and receive instant, step-by-step solutions. Features include micro-video lessons and a tutor Q&A community with over 90 million learners.

6.1.3) High School Level

Choosing the Right App

- For step-by-step solutions: Photomath, QANDA
- For structured practice and videos: Khan Academy, Mathspace
- For interactive concepts and puzzles: GeoGebra, DragonBox, Brilliant
- For deeper insights & context: MathColumn

Please see above for addresses: *Khan Academy, Photomath, GeoGebra. QUANDA, DragonBox* and consider the following:

1. *Mathspace*: **https://mathspace.co/us**

This is an adaptive learning platform for ages 7–18. It personalizes problems, gives real-time feedback, and includes hints and video tutorials.

2. *MathColumn*: **https://mathcolumn.com/**

This is a free "digital math magazine" with interactive lessons, puzzles, real-world math context, and weekly content - ideal for conceptual understanding.

3. *Brilliant*: **https://brilliant.org/**

Brilliant offers free daily problem-solving challenges and interactive courses on logic, algebra, and is an excellent catalyst to promote deeper conceptual thinking.

6.2 SCIENCE

6.2.1) Elementary Level

1. *National Geographic Kids*: **https://kids.nationalgeographic.com/**

This site explores amazing animals, countries, and science facts through colorful videos, games and articles. Great for curious minds who love nature and adventure.

2. *NASA Kids' Club*: **https://www.nasa.gov/kidsclub/**

NASA's Kids' Club lets students explore planets, astronauts, space robots, and more through fun games and facts.

3. *Mystery Science:* **https://mysteryscience.com/**

This site offers easy-to-follow science lessons with videos and hands-on activities. Topics include weather, animals, the human body, and more. (Free access for many teachers and schools.)

4. *Switch Zoo:* **https://www.switchzoo.com/**

Make your own animals and learn about real ones! Switch Zoo lets kids mix animal parts and teaches fun facts about habitats, diets, and adaptations.

5. *Science Bob:* **https://sciencebob.com/**

Science Bob shows step-by-step experiments, videos, and answers science questions.

6. *KidsHealth in the Classroom – Human Body:* **https://classroom.kidshealth.org**

This site provides kid-friendly lessons about the human body, including topics like bones, muscles, the brain, heart, and digestion. It features age-appropriate videos, printable worksheets, quizzes, and teacher guides. The content is designed to help elementary students understand how their bodies work in a fun and accessible way.

6.2.2) Middle School Level

1. *Science News Explores:* **https://www.snexplores.org**

A student-friendly site by *Science News* that makes current science events understandable for middle school readers. It covers a wide range of topics like space, health, environment, and technology and is great for reading practice and science literacy.

2. *NASA's Space Place:* **https://spaceplace.nasa.gov**

This site is full of games, hands-on activities, facts, and videos. It simplifies space science topics like planets, weather, and satellites, and is perfect for middle schoolers exploring Earth and space science.

3. *National Geographic Education:* **https://education.nationalgeographic.org**

This site offers free classroom resources, including videos, maps and lesson plans. Topics include ecosystems, climate change, geography, and conservation. High-quality visuals and content support both science and social studies learning.

4. *Science Buddies:* **https://www.sciencebuddies.org**

A go-to site for science fair projects, experiments, and **STEM** activities. Tailored to grades K–12, its middle school filter makes it easy to find age-appropriate topics in biology, chemistry, engineering, and more.

5. *CK-12 Foundation:* **https://www.ck12.org**

This free platform offers interactive science (and math) textbooks, simulations, and practice questions. Students can explore topics like forces, cells, atoms, and ecosystems through *FlexBooks* and concept-based learning.

What Are FlexBooks?

- Digital textbooks that can be edited and adapted by teachers, schools, or students,
- Cover subjects like science, math, and engineering, primarily for grades 6–12; and,
- Available online for free.

6. *TeachEngineering – Human Body Systems*: **https://www.teachengineering.org**

This site provides STEM-focused lessons that explore human anatomy in connection with engineering design. Lessons include hands-on activities about bones, muscles, and organs appropriate for middle schoolers.

7. *Innerbody:* **https://www.innerbody.com**

With interactive 3D anatomy visuals covering all major human body systems, middle school students can explore each system visually with easy-to-understand summaries and diagrams.

6.2.3) High School Level

Please see above: https://www.khanacademy.org/science, https://education.nationalgeographic.org, https://www.ck12.org/student/ and https://phet.colorado.edu

Additionally, please consider the following:

1. *NASA Science:* **https://science.nasa.gov**

This site offers up-to-date information on space exploration, astronomy, Earth science, and missions. Includes student-friendly content, images, videos, and learning resources.

2. *HHMI Biointeractive:* **https://www.biointeractive.org**

Filled with biology and life science videos, animations, and classroom activities from the Howard Hughes Medical Institute, this is great for AP Biology and deeper learning.

3. *The Physics Classroom:* **https://www.physicsclassroom.com**

This site offers simplified tutorials, animations, and practice problems for high school physics. It covers motion, forces, waves, electricity, and more.

4. *Science News for Students (Society for Science):* **https://www.snexplores.org**

This site provides current science news articles written in student-friendly language and is great for developing scientific literacy.

6.2.4) Lab/Experimental Simulations

1. *ExploreLearning Gizmos:* **https://www.explorelearning.com/**

Offering over 400 interactive math and science simulations for Grades 3–12, *Gizmos* aligns with state standards and enhance lab experiences through virtual exploration.

2. *The Concord Consortium – STEM Simulations:* **https://concord.org/**

This site shares free interactive models and simulations for science, math, and engineering. Activities cover energy, atoms, ecosystems, climate, and more. It is ideal for middle and high school, with some elementary options.

3. *Molecular Workbench:* **https://mw.concord.org/**

A specific platform from Concord Consortium focuses on high school simulations of chemistry and biology at the molecular level and includes embedded assessments and open-ended experimentation.

4. *PBS LearningMedia – Science Simulations:* **https://www.pbslearningmedia.org/subjects/science/**

Rich in video-based and interactive science simulations for K–12, *PBS LearningMedia* includes virtual labs, experiment walkthroughs, and teacher guides.

5. *Tinkercad – Circuit & Design Simulations:* **https://www.tinkercad.com/**

A free, beginner-friendly platform for Grades 4–12 allows students to simulate and build 3D models, electronics, and circuits. It is great for STEM, robotics, and engineering activities.

6.3 STEM (SCIENCE, TECHNOLOGY, ENGINEERING, MATH)

6.3.1) STEM Resources

SCIENCE: Please see the resources in 6.2.

TECHNOLOGY (and CODING)

TECHNOLOGY

1. *BrainPOP / BrainPOP Jr.:* **https://www.brainpop.com AND https://jr.brainpop.com**

These sites offer animated videos, quizzes, games, and interactive activities on a wide range of technology and digital literacy topics-such as internet safety, how computers work, and simple machines. BrainPOP Jr. is ideal for younger grades.

2. *Common Sense Education*: **https://www.commonsense.org/education**

A leading resource for teaching digital citizenship, media balance, online safety, and responsible technology use, the site includes ready-to-use lesson plans for elementary teachers.

3. *PBS LearningMedia – Technology & Engineering:* **https://www.pbslearningmedia.org/subjects/engineering-and-technology/**

With interactive lessons and videos covering real-world technology topics such as robotics, transportation, communication systems, and inventions, this site is ideal for Grades K–6.

4. *National Geographic Education – Technology & Innovation:* **https://education.nationalgeographic.org/topics/technology-and-innovation**

This site provides articles, videos, maps, and activities that connect technology to real-world applications like mapping, environmental tools, and space exploration.

5. *Exploratorium – Science & Tech Activities:* **https://www.exploratorium.edu/education**

Featuring hands-on STEM activities and digital tools, this site helps students understand the science and technology in everyday life, such as how sensors, cameras, and lenses work.

CODING

1. *Code.org:* **https://www.code.org**

A leading platform offering free, curriculum-aligned coding courses for K–12. Includes fun activities like Minecraft, Star Wars, and Dance Party to introduce block-based and text-based programming.

2. *Scratch:* **https://scratch.mit.edu**

A project of MIT, *Scratch* teaches kids (Grades 3–8) to create games, animations, and interactive stories using a visual, block-based coding language.

3. *ScratchJr:* **https://www.scratchjr.org**

Designed for younger students (Grades K–2), this simplified version of *Scratch* lets students build stories and games by snapping coding blocks together on tablets.

4. *Tynker:* **https://www.tynker.com**

This site offers self-paced coding lessons for Grades 1–12, including block coding, Python, JavaScript, and AI with themed challenges like robotics and Minecraft modding.

5. *Blockly Games:* **https://blockly.games**

A beginner-friendly coding platform using Google's *Blockly* engine is ideal for Grades 3–6 to learn programming logic through puzzles and mini-games.

6. *Kodable:* **https://www.kodable.com**

Designed for Grades K–5, *Kodable* offers engaging programming lessons and activities that teach basic coding concepts using games and problem-solving tasks.

7. *CS First by Google:* **https://csfirst.withgoogle.com**

A free computer science curriculum for Grades 4–8 using *Scratch,* it includes themed coding projects and videos, perfect for classroom or after-school use.

8. *Replit:* **https://replit.com**

Best suited for Grades 6–12, *Replit* is an online coding platform that supports real-world programming languages like Python, JavaScript, and HTML. Great for transitioning students from block to text-based coding.

ENGINEERING

1. *TeachEngineering:* **https://www.teachengineering.org**

Offering hands-on engineering design challenges and lessons for K-12, the content aligns with NGSS and STEM standards.

2. *Curiosity Machine (by Technovation):* **https://www.curiositymachine.org**

Students tackle real-world engineering challenges through design and prototyping, supported by family and mentor guides.

3. *STEMfinity*: **https://www.stemfinity.com**

This site offers project kits and lesson plans focused on engineering, robotics, and maker challenges for young learners.

MATH: Please see resource in 6.1.

6.3.2) Part Two: Encourage STEM Literacy

Mastering literacy impacts every aspect of **STEM** education. Students must learn both the **STEM** concepts involved, and *how to effectively speak about them.* Simply doing and knowing are not the full picture; the ability to describe and explain is critical in all areas of learning. For longer term promotion of **STEM** literacy, here are ideas and avenues to consider.

Reading in STEM: These resources integrate engaging **STEM** topics with reading comprehension strategies.

Epic! (Digital Library for Kids): **https://www.getepic.com/**

- Offers thousands of STEM-themed books for early readers.
- Includes read-aloud options for developing fluency.

National Geographic Kids: **https://kids.nationalgeographic.com/**

- Engaging non-fiction STEM articles with vibrant images and vocabulary support.
- Encourages critical thinking with comprehension quizzes.

Science A-Z (Part of Learning A-Z): **https://www.sciencea-z.com/**

- Provides leveled STEM books with comprehension activities.
- Includes interactive digital books and printable worksheets.

StoryBots: Ask the StoryBots: **https://www.storybots.com/**

- **STEM**-based animated stories help children understand scientific concepts.
- Offers books and videos that encourage curiosity in science and technology.

ReadWorks – STEM Article Sets: **https://www.readworks.org/**

- Provides free, high-quality **STEM** reading passages with comprehension questions.
- Aligns with different reading levels for differentiation.

Writing in STEM: These tools encourage students to write about **STEM** concepts, fostering critical thinking and communication.

WriteReader: **https://www.writereader.com/**

- A digital tool where students create their own STEM books with pictures and text.
- Supports early literacy and scientific explanations.

Mystery Science Writing Prompts: **https://mysteryscience.com/**

- Engages students with science-related "mysteries" and writing reflections.

- Promotes explanatory writing about STEM concepts.

Scholastic Story Starters: STEM Edition: **https://www.scholastic.com/teachers/story-starters/**

- Generates STEM-focused creative writing prompts.
- Encourages narrative writing with scientific themes.

NASA Kids' Club Writing Challenges: **https://www.nasa.gov/kidsclub/**

- Provides space-themed writing activities.
- Includes storytelling and scientific explanations.

PBS Kids Design Squad Writing Challenges: **https://pbskids.org/designsquad**

- Encourages students to document and explain engineering challenges.
- Develops writing skills through reflection on hands-on **STEM** projects.

Speaking and Listening in STEM: These resources develop students' oral communication skills through **STEM** discussions, presentations, and collaborative conversations.

Speaking Resources

Flip (formerly Flipgrid) – STEM Video Responses: **https://info.flip.com/**

- Allows students to record and share their STEM discoveries.

- Encourages discussion and peer feedback.

Kid-Friendly TED Talks (TED-Ed for Kids): **https://ed.ted.com/**

- Features engaging STEM-related videos with discussion prompts.
- Encourages students to share their thoughts verbally.

Wonderopolis – STEM Wonder of the Day: **https://wonderopolis.org/**

- Provides intriguing STEM questions for students to discuss and explain.
- Encourages verbal reasoning and articulation of ideas.

Skype a Scientist: **https://www.skypeascientist.com/**

- Connects classrooms with real scientists for live Q&A sessions.
- Enhances verbal interaction and questioning skills.

Mystery Doug – Video-Based Discussions: **https://mysterydoug.com/**

- Engages students with intriguing science questions and fosters classroom discussions.
- Encourages students to express their thoughts and predictions.

Listening Resources

Brains On! Science Podcast for Kids: **https://www.brainson.org/**

- Features interactive science stories and Q&A sessions.
- Encourages active listening and discussion.

WOW in the World (STEM Podcast by NPR): **https://tinkercast.com/shows/wow-in-the-world/**

- **STEM**-focused storytelling podcast that engages young learners.
- Includes discussion prompts for comprehension.

Listenwise – STEM Audio Stories: **https://listenwise.com/**

- Provides science and technology-themed audio content with comprehension activities.
- Develop active listening skills.

BBC Bitesize Science Audio Clips: **https://www.bbc.co.uk/bitesize**

- Features short, engaging science lessons in an audio format.
- Help students focus on key STEM concepts while listening.

Read-Aloud STEM Books on YouTube (STEM Read Alouds)

- Features engaging read-alouds of STEM books.
- Reinforces listening comprehension and vocabulary.

For example: **https://www.youtube.com/c/StoryTimeWithMsBecky**

Reading Mathematical Problems and Technical Texts

MathStart Books: Engaging picture books that teach math concepts (Amazon)

Epic! Digital Library: A collection of math-related **STEM** books with read-aloud options: **https://getepic.com**

Greg Tang Math: Books and interactive puzzles to improve problem-solving: **https://tangmath.com**

"*Sir Cumference*" Book Series: Fun, medieval-themed math adventures (Amazon)

6.4 AI AND LANGUAGE ARTS

Applying AI to Language Arts is like having a personal muse to inspire, or a talented editor to polish your writing. It expands students' creative boundaries and encourages narrative possibilities never before realized.

Imagine your students are writing a short story. They start with a basic plot outline and then use AI to generate ideas for character development, plot twists, or even dialogues. Different styles and narrative structures arise as AI makes the writing process more immediate and interactive. AI becomes a co-writer, enriching students' creative expression, providing suggestions and helping them see their stories from a new perspective.

Applied to reading comprehension, AI applications can automatically analyze texts and generate questions and summaries. For example, after a lesson on Shakespeare's *Macbeth*, an AI tool can generate comprehension questions that challenge students to think critically about the themes and characters.

AI also can provide summaries that help students grasp complex plots and old English prose, making the entire reading process more accessible and enjoyable. It depends on your prompt. (**Note**: Drafting effective AI prompts is covered in the next chapter).

Here are eleven (11) excellent sites for Language Arts activities.

1. *ReadWriteThink:* **https://www.readwritethink.org**

Offering free, standards-based reading and writing resources, including interactive student tools, lesson plans, and printables, this is great for Grades K-12.

2. *CommonLit:* **https://www.commonlit.org**

This site provides free, curated reading passages for Grades 3-12 with built-in questions, vocabulary help, and teacher tools and includes fiction, nonfiction, and poetry.

3. *Epic!:* **https://www.getepic.com**

A digital library with thousands of books, audiobooks, and read-to-me stories for Grades K-6. This site is great for independent reading and literacy skill development.

4. *Storyline Online:* **https://www.storylineonline.net**

Ideal for Grades 1-5 to promote listening comprehension and a love of reading, celebrated actors read children's books aloud with animations.

5. *Newsela:* **https://www.newsela.com**

Delivers current events and nonfiction texts at multiple reading levels (Grades 2-12). Articles come with quizzes and writing prompts, supporting literacy and critical thinking.

6. *Scholastic Learn at Home / Scholastic Kids:* **https://www.scholastic.com/kids**

Offering interactive stories, games, book lists, and writing prompts, The Scholastic Learn at Home portal provides structured literacy activities for Grades 1-8.

7. *NoRedInk:* **https://www.noredink.com**

A grammar and writing platform for Grades 4-12, students receive personalized exercises based on their interests to improve sentence structure, punctuation, and essay writing.

8. *Vocabulary.com:* **https://www.vocabulary.com**

For grades 3-12 students, learn vocabulary through engaging, adaptive quizzes and games, with built-in definitions and usage examples.

10. *QuillBot:* **https://quillbot.com**

QuillBot offers AI-powered writing tools including paraphrasing, grammar checking, summarization, citation generation, plagiarism detection, and tone analysis. It helps students refine and streamline their writing across various formats, web pages, documents, and browser extensions, without sacrificing their unique voice.

11. *Grammarly*: **https://www.grammarly.com**

A comprehensive AI writing assistant, *Grammarly* checks grammar, spelling, punctuation, clarity, tone, and style. Including advanced features like plagiarism detection and generative AI suggestions and available as a browser extension, desktop app, mobile keyboard, and Microsoft Office add-in, it offers real-time writing assistance across platforms.

6.5 SOCIAL STUDIES AND AI

AI applied to social studies can transform your classroom into a virtual time machine. Combining AI with the power of virtual reality (VR) tools can:

- simulate historical events, "first-hand,"
- walk student through ancient Rome: or,
- witness the signing of the Declaration of Independence. All things are possible.

This kind of learning is more than simply engaging; it also deepens understanding by allowing students to see the context of these events, the emotions of the people involved, and the complexity of the decisions they made.

Incorporating AI into social studies can change dramatically how we teach history - shaping how students perceive and interact with the past. It can turn history lessons into dynamic explorations of human experiences and broaden perspectives.

By integrating these AI tools into your curriculum, you're not just teaching history; you're inspiring historians - curious, empathetic, and connected to a global community.

6.5.1) Elementary School Level

1. *National Geographic Kids:* **https://kids.nationalgeographic.com/**

This fun, interactive site covers geography, cultures, history, and animals with videos, games, and quizzes.

2. *Ducksters: Social Studies for Kids:* **https://www.ducksters.com/history/**

These easy-to-read articles on world history, U.S. history, geography, and biographies are geared toward elementary learners.

3. *PBS Kids – Social Studies:* **https://pbskids.org/games/social-studies/**

Games and videos that teach civics, community, and U.S. history using favorite PBS characters make learning fun for kids.

4. *iCivics – Elementary:* **https://www.icivics.org/teachers/lesson-plans**

This site offers simplified civics lesson plans, games, and printable materials for younger students.

5. **Mr. Nussbaum – Social Studies: https://mrnussbaum.com/social-studies**

Enjoy these interactive activities, reading comprehension passages, and games focused on U.S. presidents, geography, and history.

6.5.2) Middle School Level

1. *iCivics:* **https://www.icivics.org/**

These engaging, gamified lessons on U.S. government, the Constitution, rights, and civic responsibility make learning fun.

2. *National Geographic Education:* **https://education.nationalgeographic.org/**

These resources, maps, videos, and lessons on geography, culture, and world history are tailored for grades 6-8.

3. *Smithsonian Learning Lab – Social Studies:* **https://learninglab.si.edu/**

Access thousands of curated resources on U.S. history, world cultures, and civic education with interactive learning tools.

4. *CIA World Factbook – Student Edition:* **https://www.cia.gov/the-world-factbook/**

This site offers country-by-country information useful for world geography and global studies projects.

5. *Teaching Tolerance (Learning for Justice):* **https://www.learningforjustice.org/classroom-resources**

This site provides lesson plans on social justice, civil rights, and identity studies appropriate for middle school learners.

6.5.3) High School Level

1. *Facing History and Ourselves*: **https://www.facinghistory.org/**

Thoughtful lessons on racism, prejudice, and historical events like the Holocaust and civil rights movement.

2. *CrashCourse – YouTube Social Studies Series:* **https://www.youtube.com/user/crashcourse/playlists**

High-quality, fast-paced video lessons on U.S. History, World History, Government, and Economics.

3. *Stanford History Education Group (SHEG):* **https://sheg.stanford.edu/**

This site uses primary sources to develop critical thinking through inquiry-based history lessons.

4. *EconEdLink (Council for Economic Education):* **https://www.econedlink.org/**

Search free lessons and resources to teach economics and personal finance to high school students.

5. *Digital History:* **https://www.digitalhistory.uh.edu/**

This site provides an extensive collection of documents, timelines, and multimedia content focused on American history.

6.6 AI IN ART: FOSTERING CREATIVITY THROUGH TECHNOLOGY

When imagining a new world of art education, AI expands the horizons of what art can be and how it's taught. AI tools allow students to input their own digital paintings and see how AI can enhance their work with surreal, imaginative details.

The use of AI in manipulating and/ or creating art also allows students to experiment with new techniques that can be complex or impossible to achieve manually. Further, AI-based painting programs can mimic the styles of historical art movements such as Impressionism or Cubism.

The intersection of AI with traditional art forms provides new avenues for innovation and creativity. For instance, students could work on a project that combines poetry and graphic design, using AI to generate visually appealing layouts of their poems.

Another exciting cross-disciplinary project can involve music and technology whereby students use AI to compose music that reflects the themes or emotions of a particular painting.

Integrating AI into art education prepares students for a world where digital literacy is intertwined with every aspect of creativity and self-expression. Through AI, we can provide a platform for students to explore, create, and innovate in ways never before realizable.

1. *Art for Kids Hub:* **https://www.artforkidshub.com/**

This fun, family-friendly site offers thousands of step-by-step drawing, painting and craft video tutorials suitable for elementary through middle school students.

2. *The Art of Education University:* **https://theartofeducation.edu/**

Offering art lesson plans, curriculum materials, and professional development for K-12 art educators, this site includes a wide range of activities, assessments, and studio projects for all grade levels.

3. *KinderArt:* **https://kinderart.com/**

A comprehensive site with free art lessons and activities for grades K-12, *KinderArt* covers drawing, painting, sculpture, and seasonal crafts, along with integration ideas for other subjects.

4. *Deep Space Sparkle:* **https://www.deepspacesparkle.com/**

Focusing primarily on elementary and middle school art education, this site offers detailed lesson plans, how-to videos, and classroom resources across a wide range of media.

5. *The Getty Museum – Education:* **https://www.getty.edu/education/**

A rich archive of visual arts resources for grades K-12 including lesson plans, image galleries, and art history connections, this is great for integrating fine arts into humanities or history classes.

6. *National Gallery of Art – NGAkids:* **https://www.nga.gov/education/kids.html**

NGAkids offers interactive online art tools and educational games for K-8, as well as printable activities and links to art collections for older students to encourage creativity while exploring famous artworks.

6.7 AI: SPORTS AND PHYSICAL EDUCATION

1. *PE Central:* **https://www.pecentral.org/**

One of the most widely used resources for physical education teachers, *PE Central* offers thousands of lesson plans, activity ideas, and assessment tools for all grade levels.

2. *OPEN Physical Education:* **https://openphysed.org/**

A free resource with high-quality physical education units and lesson plans designed for K-12 students, it includes fitness, dance, sports, and SEL (social-emotional learning) activities.

3. *SPARK PE:* **https://sparkpe.org/**

Evidence-based PE programs and curriculum resources for teachers from grades K-12, this site includes training materials, videos, and downloadable lesson plans.

4. *SHAPE America:* **https://www.shapeamerica.org/**

The national organization for health and physical educators offers standards, lesson ideas, professional development, and advocacy tools for K-12 physical education.

5. *PBS LearningMedia – Health & Physical Education:*

https://www.pbslearningmedia.org/subjects/health-and-physical-education/

PBS provides videos, interactive activities, and lesson plans related to sports, fitness, and health for elementary through high school students.

6. *GoNoodle:* **https://www.gonoodle.com/**

Especially popular with elementary schools, *GoNoodle* offers free movement and mindfulness videos designed to get kids moving during the school day; it is ideal for brain breaks or indoor recess.

7. *TeamBuildr*: **https://www.teambuildr.com/**

For coaches and players, AI is quickly transforming how we approach fitness and team sports in educational settings. AI tools can generate personalized training programs and give teams guidance in strategizing to improve team dynamics.

AI-powered fitness apps and devices can:

- track performance,
- give real-time feedback; and,
- offer encouragement and health tips.

Wearable devices monitor heart rates, count steps, and analyze posture and movement. This data is not just numbers; it's a gateway to understanding each student's physical capabilities and needs.

An AI fitness app can take this data and craft customized workout plans tailored to individual stamina and fitness goals, making exercises more effective.

Applied to teams, AI can:

- analyze game footage, providing detailed insights into team dynamics, player movements, and game strategies,
- understanding the strengths and weaknesses of each team member and the team as a whole; and,
- highlight areas where teams excel and suggest improvements.

Overall, such data promotes good habits and fosters a healthy spirit of competition, teamwork, and ultimately victory!

7

EFFECTIVE AI PROMPTING AND ACTIVITY CREATION

In this chapter, we explore how to master the art of writing effective AI search prompts that create unique activities tailored to fulfill your goals and needs.

NOTE*:* While ChatGPT may be the most recognized, here are three (3) other AI platforms you may want to explore.

- Google's *Gemini* is a conversational AI chatbot that generates human-like text responses, assists with various tasks, and provides information in a conversational manner.
- Microsoft's *Bing Chat* is integrated with their search engine and provides conversational AI capabilities, including answering questions, assisting with tasks, and generating text.
- Anthropic's *Claude* is a conversational AI model that allows users to have natural language conversations, ask questions, and receive detailed responses.

Note: Related to ChatGPT, after providing the requested information, it commonly asks if it can be altered in any way, or if you might like a PDF version or something else.

7.1 TIPS FOR CREATING EFFECTIVE SEARCH PROMPTS

Crafting effective search prompts is akin to writing an exceptional exam question. You need clarity, focus, and an understanding of what you seek to achieve. Prompting is more than simply asking AI to generate a response; it's about crafting prompts that elicit specific responses which best serve your educational goals.

We begin with a bit of history about the foundational prompting strategy when crafting effective searches, then provide useful tips.

Boolean Style of AI Input – A Brief Description

Boolean input refers to using *logical operators* like **AND, OR, NOT**, and **( ) parentheses** to *narrow or expand searches or commands* in AI systems or databases. It helps make queries more **specific** and **accurate**.

It comes from Boolean logic, a form of algebra where all values are either true (1) or false (0).

How It Works in AI or Search Queries:

AND – *Narrows* the search. Results must include **both** terms.

Example: robots AND ethics

(Returns results that contain **both** "robots" and "ethics")

OR – *Broadens* the search. Results can include **either** term.

Example: AI OR machine learning

(Returns results that include **either** term)

NOT – *Excludes* a term.

Example: self-driving cars NOT Tesla

(Finds articles about self-driving cars, but **excludes** anything with "Tesla")

( ) Parentheses – Used to group terms and clarify order.

Example: (robots OR drones) AND surveillance

(Searches for "surveillance" involving **either** "robots" or "drones")

7.1.1) Here are nine (9) tips to help you create prompts that bring out AI's best.

1. Begin with the End in Mind

Start by clearly defining what you want your students to learn. Such specificity aligns your prompts and educational objectives.

Here are two (2) examples of **explanation** and **quiz** prompts.

EXPLANATION

Note: You will find that adding limiting phrases such as, *in 100 words or less ...,* or, *provide three examples of ...* keeps AI under control. Otherwise, depending on the prompt topic, you will get far more than you need.

For **middle school** level, prompt: *for grade 5 level, in 200 words or less, describe what photosynthesis is AND identify the steps involved.*

Here Is The Result

What is Photosynthesis? (Grade 5 Level)

Photosynthesis is the process plants use to make their own food. Plants take in sunlight, water, and carbon dioxide from the air. Using the energy from sunlight, they turn these ingredients into sugar (food) and oxygen. This happens mostly in the leaves, which contain a green part called chlorophyll that helps absorb sunlight.

Here are the main steps in photosynthesis:

1. Sunlight hits the plant's leaves.
2. Chlorophyll absorbs the sunlight's energy.
3. The plant takes in carbon dioxide through tiny holes in the leaves.
4. The plant's roots absorb water from the soil.
5. Using the sun's energy, the plant changes the water and carbon dioxide into sugar (glucose) and oxygen.
6. The plant uses the sugar for energy and growth.
7. The plant releases oxygen into the air.

Photosynthesis is important because it gives plants food and produces the oxygen we breathe. Without it, people and animals wouldn't have clean air or enough food!

Let me know if you'd like this turned into a printable worksheet or mini-poster for classroom use.

For **senior students**, prompt: *for grade 9 level, in 300 words or less, describe in detail what photosynthesis is and how photosynthesis plays a role in mitigating climate change.*

Here Is The Result

Photosynthesis and Its Role in Climate Change (Grade 9 Level)

Photosynthesis is a natural process used by green plants, algae, and some bacteria to make their own food. It takes place in chloroplasts, which are found mostly in the leaves of plants. These structures contain *chlorophyll,* a green pigment that captures energy from sunlight.

During photosynthesis, plants absorb *carbon dioxide* (CO_2) from the air through tiny pores in their leaves called *stomata.* They also take in *water* (H_2O) through their roots. Using the sun's energy, they convert these ingredients into *glucose* (a type of sugar) and *oxygen* (O_2). The glucose provides energy for the plant to grow as the oxygen is released into the atmosphere.

The formula for photosynthesis is: $6CO_2 + 6H_2O + \text{sunlight} \rightarrow C_6H_{12}O_6 + 6O_2$

This process plays a major role in the Earth's carbon cycle. It helps *remove carbon dioxide,* a greenhouse gas, from the atmosphere. Too much carbon dioxide traps heat and contributes to *climate change.*

By capturing carbon dioxide, forests, grasslands, and ocean plants act as *carbon sinks.* They store carbon in their roots, trunks, and leaves, which reduces the amount of CO_2 in the air. When forests are cut down or burned, not only is this storage lost, but CO_2 is also released back into the atmosphere.

In short, photosynthesis helps fight climate change by:

- Removing carbon dioxide from the atmosphere
- Storing carbon in plants and soil
- Releasing oxygen for animals and humans to breathe

Protecting and planting more green spaces is one way we can support photosynthesis and reduce the effects of climate change.

Let me know if you'd like this turned into a **printable worksheet** or **mini-poster** for classroom use.

To show the versatility in using AI to craft prompts, another option is to expand a prompt with additional instructions to change the focus of the activity. For example, prompt:

"for grade 9 level, in 300 words or less, in a printable activity worksheet, describe in detail what photosynthesis is and how photosynthesis plays a role in mitigating climate change BUT include three (3) false facts that students must circle AND provide a separate answer key with the correct text"

Here Is The Result

Student Worksheet: Photosynthesis and Climate Change

Name: ______________________ **Date:** ___________

Instructions: Read the passage carefully. Circle three false facts that are not true. Be ready to explain why they are incorrect.

Photosynthesis and Climate Change

Photosynthesis is the process plants use to make food by converting sunlight into energy. This takes place mostly in the *roots* of plants, where chlorophyll captures sunlight. Plants absorb *oxygen* from the air and water from the soil. Then, using energy from the sun, they convert those ingredients into glucose (a type of sugar) and carbon dioxide.

The correct formula for photosynthesis is: $6CO_2 + 6H_2O +$ sunlight $\rightarrow C_6H_{12}O_6 + 6O_2$

Photosynthesis is important because it reduces the amount of carbon dioxide (CO_2) in the atmosphere. CO_2 is a greenhouse gas that traps heat and contributes to climate change. By absorbing it, plants help cool the Earth. Forests and oceans are major carbon sinks that store carbon for long periods.

When we cut down trees, we increase the planet's ability to store carbon. Protecting plants and green areas helps reduce the effects of climate change and supports the oxygen cycle, which all living things rely on.

Answer Key (Teacher Use)

Corrected Text with False Facts Identified:

1. False Fact: *"Photosynthesis takes place mostly in the roots."*
2. Correction: It takes place mostly in the *leaves,* where chloroplasts are found.
3. False Fact: *"Plants absorb oxygen from the air..."*
4. Correction: Plants absorb *carbon dioxide* from the air, not oxygen.
5. False Fact: *"When we cut down trees, we increase the planet's ability to store carbon."*
6. Correction: Cutting down trees *reduces* the ability to store carbon.

The ability to craft unique, focused activities that directly align with your needs is limitless.

QUIZ

Note: As evidenced in the Grade 9 prompt above, including the phrase, *"in a printable activity worksheet"* to anything creates a name, date and instruction setup at the top of the page.

Here is an example of a quiz prompt: *for grade 2 level, in a printable activity worksheet, create a 6 question, multiple choice quiz about the life cycle of a butterfly AND provide a separate answer key.*

Here Is The Result

Life Cycle of a Butterfly

Name: ______________________ **Date:** ___________

Instructions: Read each question carefully. Circle the letter of the correct answer.

1. What is the first stage in the life cycle of a butterfly?
   A. Caterpillar
   B. Egg
   C. Chrysalis
2. What hatches from the egg?
   A. A baby butterfly
   B. A bee
   C. A caterpillar
3. What does a caterpillar mostly do?
   A. Fly around
   B. Eat and grow
   C. Sleep all day
4. What is the hard case the caterpillar forms called?
   A. Nest
   B. Cocoon
   C. Chrysalis
5. What happens inside the chrysalis?
   A. The caterpillar turns into a butterfly
   B. The caterpillar takes a nap
   C. The caterpillar builds a house

6. What is the last stage of the butterfly's life cycle?
   A. Egg
   B. Caterpillar
   C. Adult butterfly

Answer Key – Life Cycle of a Butterfly Quiz

1. B. Egg
2. C. A caterpillar
3. B. Eat and grow
4. C. Chrysalis
5. A. The caterpillar turns into a butterfly
6. C. Adult butterfly

2. Keep It Clear and Concise

AI thrives on specificity. The clearer and more precise your prompt, the better AI can provide relevant and accurate content. Avoid ambiguity. Use direct and specific questions or statements.

For example, instead of prompting, *provide information about the planets,* prompt: *for grade 11 level, identify three characteristics of each planet in our solar system.* Such specificity narrows AI's focus and increases satisfactory responses.

Here Is The Result

Here is a Grade 11-level chart listing three key characteristics of each planet in our solar system, focusing on advanced concepts such as composition, atmospheric conditions, and orbital behavior:

☿ Mercury

1. Highly dense due to a large iron-rich core, comprising about 85% of its radius.
2. Virtually no atmosphere, leading to extreme temperature fluctuations (–180°C to 430°C).
3. Shortest orbital period (88 Earth days); experiences 3 days every 2 years due to slow rotation.

♀ Venus

1. Thick carbon dioxide atmosphere with intense greenhouse effect—surface temperature ~465°C.
2. Retrograde rotation—rotates east to west, opposite of most planets.
3. Surface pressure is about 92 times Earth's, equivalent to being 900 meters underwater.

🌍 Earth

1. Only planet with liquid water on the surface in all three phases—solid, liquid, and gas.
2. Supports life due to stable climate, magnetic field, and protective ozone layer.
3. Composed of tectonic plates, making it geologically active with quakes and volcanic activity.

♂ Mars

1. Thin CO_2 atmosphere; unable to retain heat—average temperature is –63°C.
2. Features Olympus Mons (tallest volcano) and Valles Marineris (deep canyon system).

3. Evidence of ancient water flows; polar ice caps contain water and dry ice.

♃ Jupiter

1. Composed mainly of hydrogen and helium; likely has a metallic hydrogen interior.
2. Known for its Great Red Spot, a persistent anticyclonic storm.
3. Possesses a strong magnetic field and emits more heat than it receives from the Sun.

♄ Saturn

1. Distinguished by its complex ring system made of ice particles and rocky debris.
2. Low density—less than water—could float in a sufficiently large body of water.
3. Titan, its largest moon, has a dense nitrogen atmosphere and methane lakes.

♅ Uranus

1. Extreme axial tilt (98°)—it essentially rolls around the Sun on its side.
2. Atmosphere contains methane, giving it a blue-green color.
3. Classified as an ice giant, with water, ammonia, and methane in its mantle.

♆ Neptune

1. Strongest recorded winds in the solar system—up to 2,100 km/h.
2. Deep blue due to atmospheric methane and unknown chromophores.
3. Has a retrograde moon, Triton, thought to be a captured Kuiper Belt object.

Let me know if you'd like this in printable chart format, worksheet with questions, or flashcards for study purposes.

3. When To Use Open-Ended Questions

While specificity is crucial, prompts can also impact the richness of the output. Open-ended questions encourage more comprehensive and exploratory responses. These are particularly useful in creating discussion-based activities or creative writing assignments.

For example, prompt: *for grade 3 level, in 150 words or less, create a conversation between an eagle and a camel about where they live*

ChatGPT said:

Eagle: Hello, Camel! It's so hot down there in the desert. How do you live in such a dry place?

Camel: Hi, Eagle! I've got thick eyelashes, wide feet, and I store fat in my hump. The desert is my home. What about you?

Eagle: I live high in the mountains and fly over forests. I build my nest on cliffs and love the cool breeze up here.

Camel: Wow, I'd get dizzy up there! But I guess you need space to soar.

Eagle: And you need shade and water to stay cool. We're so different, but both built for our homes!

Camel: Exactly! Nature gave us what we need. Isn't that amazing?

Eagle: It sure is. Stay strong in the sand, friend!

Camel: And you keep flying high, Eagle!

Not only is this fun, it blends perfectly with a class about ecosystems. The activity's possibilities are endless. After sharing the above, you can assign animal pairs to students and they can write and share their own conversations.

4. Context is Key

Sometimes, providing background in your prompt can help AI understand the scope and depth of what you want. This is especially useful when dealing with complex subjects that can be interpreted in various ways.

Adding context helps narrow down AI's responses to align more closely with the specific aspect of the topic you're teaching.

For example, prompt: *In 1929, the USA stock market crashed and the Great Depression began. Identify five ways the common person adapted to this unprecedented economic catastrophe in 300 words or less for grade 6 level.*

Then, switch out *"common person"* with *"upper class" or "businesses"* and re-prompt. The differing replies can lead to great class discussions.

5. Adjust the Complexity

The age and proficiency level of your students should always guide your thinking when crafting prompts. The simpler, or more complex, your word choice, the simpler or more complex the prompt result. Also, providing specific age ranges, i.e., *For Grades 3-6 ...*, or tone style, i.e., *In a friendly tone ...*, impacts the prompt result.

6. Iterate and Improve

The first prompt might not always be perfect, and that's okay. Test different phrasing or structures in your prompts. Observe which ones produce the most useful responses in meeting your educational goals. This continuous refinement turns good prompts into great ones.

7. Leverage Keywords

Identify key terms in your lesson plans, then use them in your prompt. This increases effectiveness and accuracy. If you are discussing geological formations, depending on your goal, include terms like *sedimentary rocks, erosion,* or *plate tectonics.*

8. Avoid Leading Questions

While it's important to be specific, remember that one of your aims is to elicit unbiased, balanced AI responses that offer a broad perspective, especially on subjective topics. This is particularly crucial in subjects like history or literature where multiple viewpoints should be considered. Be careful your prompts don't inadvertently lead AI to offer biased responses.

Biased Prompt Example: *Why are electric cars better than gas cars?*

Problem: This prompt assumes **electric cars are better**, which **biases the AI** into giving only **positive responses** about electric cars and possibly negative ones about gas cars.

Unbiased Version: *What are the advantages and disadvantages of electric cars compared to gas cars?*

Why it's better: This version encourages the AI to present **both sides fairly**, helping the user understand **strengths and weaknesses** of each option without assuming superiority.

This kind of careful wording is especially important when using AI for **research, education, or decision-making**, to ensure **balanced and accurate** responses.

9. Interactive Prompts

To make the learning interactive, consider prompts that require actions or decisions from students. Prompts that involve role-playing or decision-making can turn AI interactions into dynamic educational activities.

For example, prompt: *for grade 7 level, in 300 words or less, create a printable activity between a taxi driver and a bus driver by generating a conversation about how their jobs are alike and different*

Students role play the result, adding more information as they see fit. Have pairs present their conversation to the class.

As a template, simply switch out the job pairings and re-prompt. In a class of thirty, do this 15x and you have a classroom full of fun engagement.

A Checklist for Crafting Effective AI Prompts

To help you implement these tips, here's a handy checklist that you can refer to while designing your AI search prompts:

1. **Define the Learning Objective**: What should students learn from this AI interaction?
2. **Clarity and Precision**: Is the prompt specific and clear?
3. **Open-Ended vs. Closed Questions**: Does the prompt encourage detailed responses?
4. **Contextual Information**: Have I provided enough context for the AI to generate relevant content?
5. **Complexity Adjustment**: Is the prompt appropriately complex for my students' level?
6. **Repetitive Improvement**: How can I refine the prompt based on AI performance?
7. **Keyword Inclusion**: Have I included relevant keywords to focus the AI's response?
8. **Unbiased Questioning**: Is the prompt designed to elicit unbiased responses?
9. **Interactive Elements**: Does the prompt include elements that require student interaction?

This checklist can serve as a quick reference to ensure that every AI prompt you create is well-crafted, leading to effective, engaging responses. Whether you're discussing Shakespeare or the laws of motion, these strategies enhance AI's capabilities to meet your educational goals, making every lesson an exciting adventure in learning.

8

AI ACTIVITY & LESSON PLANS SAMPLES

As **Power Up Your Classroom With AI** nears its' end, I would like to provide a wide variety of prompted activities and lesson plans to illuminate AI's possibilities. Enjoy!

8.1 AI ACTIVITY SAMPLES

Weather by Class Levels

Let's explore the results of the same prompt topic tailored to different grade levels: 2-5, 6-9, and 10-12.

Prompt: *In 150 words or less, describe 3 activities for Grades 2-5 students about winter weather*

Here Is The Result

1. **Create Your Own Snowflake**: Students use basic craft materials to create paper snowflakes. This activity not only sparks creativity but also serves as a hands-on lesson in

symmetry and geometry. As each student unfolds their snowflake, it's a wonderful moment to discuss why no two snowflakes are exactly alike, introducing concepts of uniqueness in natural formations.

2. **Winter Weather Diary**: Over a week, students maintain a diary documenting the weather. They draw pictures and write short descriptions of what they see and feel. This continuous observation encourages attentiveness and note-taking skills. At the week's end, review the entries as a class and discuss the patterns observed, bringing in concepts of weather changes and winter characteristics.
3. **Freezing Point Experiment**: Here, students learn about the freezing point through a simple experiment using water, salt, and ice. They predict what will happen when salt is added to ice and then conduct the experiment to observe the melting speed. This experiment is a fantastic gateway to discussions about how salt is used in winter road safety in real life.

Prompt: *In 150 words or less, describe 3 activities for Grades 6 -9 students about winter weather*

Here Is The Result

1. **Interactive Weather Map Project**: Students use online tools to create interactive weather maps that track a winter storm. They research weather patterns and use digital tools to represent the storm's path visually. This activity boosts tech skills and deepens understanding of meteorological patterns and the science of storm formation.

2. **Debate on Climate Change and Winter Weather**: Organize a debate where students discuss how climate change is impacting winter weather globally. This activity encourages critical thinking and research skills. Students must gather evidence to back their arguments. It also heightens awareness of global environmental issues and fosters public speaking skills.
3. **Write a Winter Weather Report**: Students write and record their own winter weather reports using video or audio tools. They research current weather data, write a script, and present it as if they were meteorologists. This integrates science with technology and arts, enhancing research skills and creativity while boosting confidence in public speaking.

Here, the focus shifts towards a deeper scientific understanding of winter phenomena, combined with technology integration.

Prompt: *In 150 words or less, describe 3 activities for Grades 10 -12 students about winter weather.*

Here Is The Result

1. **Scientific Research Paper on the Effects of Winter Weather**: Students choose a specific aspect of winter weather, such as its impact on agriculture or urban planning, and write a detailed research paper. This task enhances academic research skills and deepens understanding of the interconnectedness of weather phenomena and societal impacts.
2. **Develop a Mobile App for Winter Survival Tips**: In this project, students design a concept for a mobile app that offers winter survival tips, including first-aid, warming

shelters, and emergency contacts. This activity merges computer science with practical life skills and encourages students to think innovatively about how technology can serve community needs.

3. **Field Study of Local Ecosystems in Winter**: Conduct a field study where students observe how local plant and animal life adapt to winter conditions. Following the observations, students prepare a presentation or infographic that summarizes their findings, linking back to lessons on ecosystems and adaptation. This project brings biology to life and highlights the importance of biodiversity and environmental stewardship.

These activities are designed to foster advanced analytical skills and encourage explorations that connect broader environmental issues with scientific concepts.

The Properties of Light

Prompt: *create a grade 4 level, 12 question game that teaches students about the properties of light - reflect and refract with game cards included for classroom use AND a separate answer key*

Here Is The Result

(**Note**: I do sometimes alter the original result. For example, it used "*game cards*" and I switched it to "*card strips*" among some other presentational aspects).

Worksheet Game: "Light Quest – Reflect or Refract?"

Instructions:

1. Cut out the 12 game strips below and place them face down.
2. In pairs or small groups, students take turns selecting one.
3. Read the question aloud.
4. Choose the correct answer from the multiple choices.
5. Keep the card if you get it right.
6. The player with the most cards at the end wins!

Game Strips

1Q: What happens when light bounces off a mirror?

A) Refraction B) Reflection C) Absorption

2Q: What happens when light bends while passing through water?

A) Reflection B) Shadow C) Refraction

3Q: Which of these objects reflects light best?

A) Black shirt B) Shiny metal C) Cardboard

4Q: When light bends, it's called:

A) Reflection B) Refraction C) Shading

5Q: A mirror causes light to:

A) Go through B) Bounce off C) Disappear

6Q: A magnifying glass bends light. What is this called?

A) Refraction B) Dissolving C) Reflection

7Q: What tool do we use to see our reflection?

A) Flashlight B) Mirror C) Glass

8Q: What material would bend light the most?

A) Air B) Clear water C) Thick glass

9Q: You see your face in a shiny spoon because of:

A) Sound B) Friction C) Reflection

10Q: What happens when light passes from air into water?

A) It disappears B) It bends C) It reflects

11Q: Sunglasses help block light by:

A) Reflecting it B) Refracting it C) Absorbing it

12Q: Why do pencils look bent in a glass of water?

A) Reflection B) Refraction C) Vibration

Answer Key: Light Quest – Reflect or Refract?

Card # Answer

1. B) Reflection
2. C) Refraction
3. B) Shiny metal
4. B) Refraction
5. B) Bounce off
6. A) Refraction
7. B) Mirror
8. C) Thick glass
9. C) Reflection
10. B) It bends
11. C) Absorbing it

12. B) Refraction

The game encourages teamwork and problem-solving, while teaching how light behaves when it bounces off surfaces (reflection) or bends through materials (refraction).

The Factor Frenzy Game

prompt: *in 150 words or less, create a Grade 4 level game that introduces factors, multiples, and prime numbers*

Here Is The Result:

(Note: I added the materials section).

Materials: Create the required number of index card sets. Each card has a single number from 1-50.

How to Play:

Each player gets a number card (between 1–50). Take turns rolling a die. The number you roll tells you what to do:

1: Name **two factors** of your number.

2: Name a **multiple** of your number.

3: Say if your number is **prime or composite**.

4: Name a **factor pair** (like $2 \times 6 = 12$).

5: Name a **prime number** less than your number.

6: Trade cards with another player.

Correct answers = 1 point. First to 10 points wins!

This fast-paced game helps students explore **factors, multiples**, and **prime numbers** in a fun, active way. Use number cards, a die, and a score sheet to play.

All About Macbeth

prompt: *In 200 words or less, for Grade 11 level, create a 5 question, multiple choice quiz about Macbeth AND on a separate page, provide the answers and explanations*

Here Is The Result:

Macbeth Multiple Choice Quiz (Grade 11)

1. **Who is the first character to be killed by Macbeth?**
   a. Banquo
   b. Duncan
   c. Macduff
   d. Lady Macbeth
2. **What do the witches predict for Banquo?**
   a. He will be king
   b. His children will be kings
   c. He will die in battle
   d. He will betray Macbeth
3. **Why does Lady Macbeth want her husband to kill Duncan?**
   a. To prove his loyalty
   b. To become queen
   c. To avenge a wrong
   d. She loves violence

4. **What vision does Macbeth see before he murders Duncan?**
   a. A bloody crown
   b. A floating dagger
   c. A ghostly figure
   d. Banquo's ghost
5. **Which title does Macbeth receive after his victory at the beginning of the play?**
   a. Thane of Fife
   b. Thane of Glamis
   c. Thane of Cawdor
   d. Prince of Cumberland

Answer Key and Explanations

1. **b) Duncan**
   - Macbeth murders King Duncan to fulfill the witches' prophecy and seize the throne.
2. **b) His children will be kings**
   - The witches predict that Banquo's descendants will become kings, not Banquo himself.
3. **b) To become queen**
   - Lady Macbeth pushes Macbeth to kill Duncan so they can rule as king and queen.
4. **b) A floating dagger**
   - Before killing Duncan, Macbeth hallucinates a dagger leading him toward the act.
5. **c) Thane of Cawdor**
   - After his victory, Macbeth is awarded the title Thane of Cawdor by King Duncan.

Prompting Images: These prompts provide images that are great for cutting out and using for games or activities related to ecosystems, predator-prey relationships, and the food cycle.

Here Is The Result

Prompt: *create a simple 3x2 black & white image in a savannah environment of a predator (jaguar) chasing its prey (cottontail rabbit)*

Prompt: *on a single sheet of paper, provide 3 simple, basic, 5x3 black & white images of a polar bear, a seal and a salmon* (**Note**: I shrunk the image due to space).

The Atmosphere Above Us

Prompt: *for grade 6 level, create a printable worksheet activity with a brief topic introduction then mix up and provide three (3) facts about each layer of the atmosphere AND a separate answer sheet*

Here Is The Result

Grade 6 Science Worksheet: Layers of the Atmosphere

Name: ______________________ **Date:** ___________

Topic Introduction:

Earth's atmosphere is made up of layers, each with different characteristics. These layers help protect us, support life, and control weather and climate. The five main layers - **troposphere, stratosphere, mesosphere, thermosphere,** and **exosphere** - have unique roles and features. Below are mixed-up facts. Your job is to match each fact to the correct layer!

Instructions:

Read the facts below. Then, **write the correct layer name** (Troposphere, Stratosphere, Mesosphere, Thermosphere, or Exosphere) beside each fact.

Mixed-Up Facts:

1. __________ This layer contains the ozone layer that absorbs harmful UV radiation.
2. __________ This is the coldest layer of the atmosphere.
3. __________ This layer is where weather events like rain and snow occur.
4. __________ This layer extends into space and has very thin air.
5. __________ Satellites orbit the Earth in this layer.
6. __________ This layer is where most commercial airplanes fly.
7. __________ This is the layer closest to Earth's surface.
8. __________ The Northern and Southern Lights (auroras) occur here.

9. __________ This layer can reach extremely high temperatures due to solar activity.
10. __________ Meteors often burn up in this layer before reaching the Earth.
11. __________ This layer has the greatest air pressure.
12. __________ This layer gradually fades into outer space.

Challenge Question: Which layer protects us from harmful ultraviolet rays?

Answer: ______________________________

Answer Sheet – Layers of the Atmosphere

1. Stratosphere
2. Mesosphere
3. Troposphere
4. Exosphere
5. Thermosphere
6. Stratosphere
7. Troposphere
8. Thermosphere
9. Thermosphere
10. Mesosphere
11. Troposphere
12. Exosphere

Challenge Answer: Stratosphere

Would you like this formatted into a printable PDF or Word layout with a title and boxed sections?

Understanding ***Lord of the Flies*** **by William Golding**

prompt: *select a complex paragraph from the novel, Lord of the Flies by William Golding AND re-write it in an easier to understand format*

Here Is The Result

Original Complex Paragraph (Chapter 9 – "A View to a Death"):

"Softly, surrounded by a fringe of inquisitive bright creatures, itself a silver shape beneath the steadfast constellation, Simon's body rested on the sand. The line of his cheek silvered and the turn of his shoulder became sculptured marble. The strange attendant creatures, with their fiery eyes and trailing vapours, busied themselves round his head. The body lifted a fraction of an inch from the sand and a bubble of air escaped from the mouth. The lines of his body were softened. Once the wind passed over the body and lifted the fringe of hair, so that the cheeks were covered softly with silver. Then the sea breathed again in a long, slow sigh, the water dressed Simon's coarse hair with brightness. The line of his body disappeared, softened and moved back again. The strange attendant creatures with their fiery eyes and trailing vapours busied themselves round his head. The great wave of the tide moved further along the island and the water lifted gently and deposited the body further along the shore."

Simplified/Easier Version:

Simon's body lay quietly on the sand, surrounded by glowing sea creatures. In the moonlight, his skin looked silver, and his body seemed like a marble statue. The small glowing creatures hovered around his head. A small bubble of air came from his mouth, and his body slowly began to drift. The wind moved his hair gently, making his face shimmer. The sea came in with a long wave, brushing his hair with light and slowly pulling his body away. The glowing sea creatures stayed close to him as the tide carried him further down the shore.

Would you like comprehension questions or a worksheet based on this passage?

Prompt: *for grade 6 level STEM study, create a printable worksheet activity that identifies 6 types of engineers AND create a 15, True/ False statement activity about the jobs each type does AND provide a separate answer key with explanations for false answers*

Here Is The Result

Grade 6 STEM Engineering Worksheet

Topic: What Do Engineers Do?

Name: ______________________ **Date:** ___________

Engineers solve problems and build things that make our lives better. There are many types of engineers, each with their own special job. Let's learn about six of them!

Six Types of Engineers:

1. **Civil Engineer** – Designs and builds roads, bridges, and buildings.
2. **Mechanical Engineer** – Designs machines and engines that move or work.
3. **Electrical Engineer** – Works with electricity and designs things like circuits, power systems, and electronics.
4. **Chemical Engineer** – Uses chemistry to create useful products like medicine, food, and fuel.

5. **Environmental Engineer** – Helps protect nature by solving pollution and waste problems.
6. **Software Engineer** – Designs apps, websites, and programs by writing code.

Activity: True or False?

Write **T** for True or **F** for False next to each sentence.

1. __________ A civil engineer builds bridges and highways.
2. __________ A software engineer writes computer code to make apps.
3. __________ Electrical engineers work only with water systems.
4. __________ A chemical engineer might help design new medicine.
5. __________ Civil engineers work in forests studying animals.
6. __________ Mechanical engineers help design car engines.
7. __________ Environmental engineers clean up pollution.
8. __________ Software engineers fix broken power lines.
9. __________ Chemical engineers never work in factories.
10. __________ Mechanical engineers can work on robots.
11. __________ Electrical engineers help design power grids.
12. __________ A civil engineer might design a school building.
13. __________ Environmental engineers design computer games.
14. __________ Software engineers help make video games.
15. __________ Electrical engineers may design cell phone parts.

Answer Key & Explanations:

1. **T** – Civil engineers build bridges and highways.
2. **T** – Software engineers create apps using code.
3. **F** – Electrical engineers work with electricity, not water.
    - ➤ *Water systems are for environmental or civil engineers.*
4. **T** – Chemical engineers often help make new medicines.
5. **F** – Civil engineers work on structures, not animal studies.
    - ➤ *Studying animals is a job for biologists, not civil engineers.*
6. **T** – Mechanical engineers design machines like car engines.
7. **T** – Environmental engineers help reduce pollution.
8. **F** – Fixing power lines is usually done by electricians, not software engineers.
    - ➤ *Software engineers work with code, not wires.*
9. **F** – Chemical engineers often work in factories.
    - ➤ *Factories are common places for chemical engineers to apply processes.*
10. **T** – Mechanical engineers can work on building and designing robots.
11. **T** – Electrical engineers design power systems like grids.
12. **T** – Civil engineers design buildings such as schools.
13. **F** – Environmental engineers don't make computer games.
    - ➤ *Game design is for software engineers, not environmental engineers.*
14. **T** – Software engineers can help create video games.
15. **T** – Electrical engineers often design electronics, including phone parts.

Would you like this formatted into a printable PDF with fun icons and borders?

8.2 SAMPLE LESSON PLANS

1. AI Created Lesson Plans (5 each) for Elementary, Middle & High School Students

Elementary School Students: prompt: *in 1000 words or less, select 5 historical events and 5 corresponding lesson plans for Grade 3*

Result

Here are five (5) historical events paired with lesson plans suitable for elementary students, particularly for a Grade 3 audience. Each lesson plan includes a brief description of activities, objectives, and learning outcome

The First Moon Landing (1969)

Event Overview:

On July 20, 1969, astronauts Neil Armstrong and Buzz Aldrin became the first humans to walk on the Moon, marking a monumental achievement in space exploration.

A Giant Leap for Mankind

- **Objective:** Students will learn about the Moon landing and understand the significance of space exploration.
- **Activity:**
    - **Introduction:** Begin with a short video (*see www.YouTube.com*) or slideshow showing the Apollo 11 mission and astronauts walking on the Moon.
    - **Discussion:** Talk about why the Moon landing was important and how it represented human ingenuity and curiosity.

 - **Hands-On Activity:** Students create a model of the Moon's surface using clay and small toy astronauts to simulate the landing.
 - **Creative Writing:** Students write a short story or a diary entry from the perspective of an astronaut landing on the Moon.
- **Learning Outcome:** Students will understand the historical significance of the Moon landing and engage creatively with the concept of space exploration.

The Signing of the Declaration of Independence (1776)

Event Overview:

On July 4, 1776, the Continental Congress adopted the Declaration of Independence, declaring the 13 American colonies independent from British rule.

Birth of a Nation

- **Objective:** Students will learn about the Declaration of Independence and its role in American history.
- **Activity:**
  - **Introduction:** Read a simplified version of the Declaration of Independence and explain the concept of independence.
  - **Discussion:** Talk about why the colonists wanted to be independent and what freedom means.
  - **Art Project:** Students create their own "Declaration of Independence" by writing down things they would want in their own country, decorating it with symbols like the American flag and an eagle.

    - **Role-Playing:** Students act out a debate between the colonists and British representatives, discussing the reasons for seeking independence.
- **Learning Outcome:** Students will gain an understanding of the importance of independence and democratic principles.

The Invention of the Printing Press (1440)

Event Overview:

Johannes Gutenberg's invention of the printing press in the 15th century revolutionized the way information was shared, leading to widespread literacy and the dissemination of knowledge.

Spreading the Word

- **Objective:** Students will learn about the printing press and its impact on communication and knowledge.
- **Activity:**
    - **Introduction:** Show images of early books and explain how books were made before and after the printing press.
    - **Discussion:** Talk about how the printing press made books more accessible and how that changed society.
    - **Hands-On Activity:** Students create their own "printing press" by carving simple designs into potatoes and stamping them onto paper to create a repetitive pattern, mimicking the idea of mass production.
    - **Reading Activity:** Provide students with short stories or fables that could have been spread thanks to the printing press and discuss their messages.

- **Learning Outcome:** Students will understand the significance of the printing press in the spread of knowledge and literacy.

The Civil Rights Movement (1950s - 1960s)

Event Overview:

The Civil Rights Movement was a struggle for social justice that aimed to end racial segregation and discrimination against African Americans, leading to significant legal and societal changes in the United States.

Lesson Plan: "Standing Up for Equality"

- **Objective:** Students will learn about the Civil Rights Movement and the importance of equality and justice.
- **Activity:**
    - **Introduction:** Share a story or video about key figures like Martin Luther King Jr. and Rosa Parks.
    - **Discussion:** Talk about what equality means and why it's important for everyone to be treated fairly.
    - **Art Project:** Students create posters promoting equality, using slogans like "Everyone is Equal" or "Stand Up for What's Right."
    - **Role-Playing:** Students reenact the Montgomery Bus Boycott, with some students playing the roles of bus riders, drivers, and civil rights activists to understand the impact of peaceful protest.
- **Learning Outcome:** Students will develop a basic understanding of the Civil Rights Movement and the ongoing importance of equality and justice in society.

The Wright Brothers' First Flight (1903)

Event Overview:

On December 17, 1903, the Wright brothers, Orville and Wilbur, made the first successful powered flight in Kitty Hawk, North Carolina, paving the way for modern aviation.

The Dream of Flight

- **Objective:** Students will learn about the Wright brothers and the invention of the airplane, exploring the concept of innovation.
- **Activity:**
  - **Introduction:** Show a video or images of the Wright brothers' first flight and explain how they made their dream of flying a reality.
  - **Discussion:** Talk about how innovation can change the world and the importance of perseverance.
  - **Hands-On Activity:** Students build simple paper airplanes and experiment with different designs to see which flies the furthest.
  - **Creative Writing:** Students write about where they would go if they could fly anywhere in the world, imagining what they would see and do.
- **Learning Outcome:** Students will appreciate the impact of the Wright brothers' invention and understand the value of creativity and innovation.

These lesson plans are designed to engage elementary students by combining historical knowledge with interactive activities. Each event is simplified to suit a Grade 3 level, ensuring that students learn about significant historical moments and connect with them on a personal level through creative and hands-on activities.

Middle School Students: Prompt: *in 1000 words or less, select 5 scientific experiments related to the atmosphere and 5 corresponding lesson plans for Grade 7*

Result

The atmosphere is a fascinating and complex layer that surrounds our planet, playing a critical role in sustaining life on Earth. Understanding the atmosphere is essential for middle school students as it connects to various scientific concepts, such as weather, climate, and the water cycle. Below are five (5) scientific experiments related to the atmosphere, along with corresponding lesson plans designed to engage middle school students.

Creating a Model of the Water Cycle

Objective:

Students will understand the water cycle's components (evaporation, condensation, and precipitation) by creating a simple model.

Materials:

- Ziplock bag
- Water
- Food coloring (optional)
- Tape
- Permanent marker

Procedure:

1. Fill the bottom of the ziplock bag with a small amount of water. Add a few drops of food coloring if desired.
2. Seal the bag and tape it to a sunny window.
3. Use the marker to label different parts of the bag as "Ocean," "Clouds," and "Rain."
4. Observe over a few days how the water evaporates, condenses, and precipitates inside the bag.

Lesson Plan:

1. **Introduction (10 minutes):** Begin with a discussion about the water cycle and its importance to Earth's atmosphere.
2. **Activity (20 minutes):** Guide students through creating their water cycle models.
3. **Observation (10 minutes):** Have students observe and record changes in their bags over several days, discussing the stages of the water cycle.
4. **Conclusion (10 minutes):** Wrap up by reviewing the water cycle stages and connecting them to real-world weather patterns.

Air Pressure Demonstration Using a Balloon and Bottle

Objective: Students will learn about air pressure and how it affects the atmosphere.

Materials:

- Plastic bottle
- Balloon
- Hot water
- Cold water

Procedure:

1. Place the balloon over the mouth of an empty plastic bottle.
2. Submerge the bottle in hot water and observe the balloon.
3. Next, place the bottle in cold water and observe the changes.
4. Discuss how the balloon inflates and deflates based on the changes in air pressure inside the bottle.

Lesson Plan:

1. **Introduction (10 minutes):** Discuss the concept of air pressure and its role in the atmosphere.
2. **Activity (20 minutes):** Perform the experiment, guiding students through each step.
3. **Discussion (10 minutes):** Encourage students to explain why the balloon reacts differently in hot and cold water.
4. **Conclusion (10 minutes):** Relate the experiment to atmospheric phenomena like wind and weather changes.

Experiment 3: Measuring Wind Speed with a DIY Anemometer

Objective:

Students will construct a simple anemometer to measure wind speed and learn how wind is generated in the atmosphere.

Materials:

- 4 small paper cups
- 2 straws
- Pencil with eraser
- Push pin
- Scissors
- Tape

Procedure:

1. Cut the rims of two cups to make them lighter.
2. Tape the cups to the ends of two straws.
3. Cross the straws and tape them together in the middle.
4. Push the pin through the straws' intersection into the pencil eraser, allowing the structure to spin.
5. Test the anemometer by blowing on it or taking it outside to measure wind speed.

Lesson Plan:

1. **Introduction (10 minutes):** Introduce the concept of wind and its role in weather patterns.
2. **Activity (25 minutes):** Guide students through building their anemometers and testing them.
3. **Observation (5 minutes):** Discuss how wind speed is measured and its importance in meteorology.

4. **Conclusion (10 minutes):** Connect the experiment to larger atmospheric concepts, such as wind formation and weather forecasting.

Experiment 4: Cloud Formation in a Jar

Objective:

Students will explore the conditions required for cloud formation in the atmosphere.

Materials:

- Jar with a lid
- Hot water
- Ice cubes
- Aerosol spray (hairspray or air freshener)

Procedure:

1. Pour a small amount of hot water into the jar.
2. Spray a quick burst of aerosol into the jar and quickly cover it with the lid.
3. Place ice cubes on top of the lid.
4. Observe as a cloud forms inside the jar.

Lesson Plan:

1. **Introduction (10 minutes):** Discuss how clouds form in the atmosphere and the role of temperature and particles.
2. **Activity (20 minutes):** Guide students through the cloud formation experiment.

3. **Observation (10 minutes):** Have students observe and describe the cloud formation in their jars.
4. **Conclusion (10 minutes):** Relate the experiment to real-world cloud formation and weather prediction.

Experiment 5: Simulating the Greenhouse Effect

Objective: Students will learn about the greenhouse effect and its impact on Earth's atmosphere.

Materials:

- 2 clear plastic bottles
- 2 thermometers
- Plastic wrap
- Tape
- Lamp with an incandescent bulb

Procedure:

1. Place a thermometer inside each bottle.
2. Cover the mouth of one bottle with plastic wrap and seal it with tape.
3. Place both bottles under a lamp and turn on the light.
4. After a few minutes, record the temperature in both bottles and compare.

Lesson Plan:

1. **Introduction (10 minutes):** Explain the greenhouse effect and its importance in regulating Earth's temperature.

2. **Activity (20 minutes):** Guide students through the experiment, explaining each step.
3. **Discussion (15 minutes):** Discuss why the temperature is higher in the covered bottle and relate it to the real greenhouse effect.
4. **Conclusion (5 minutes):** Discuss the implications of the greenhouse effect on global warming and climate change.

These five (5) experiments provide middle school students with hands-on learning experiences that enhance their understanding of the atmosphere. Each experiment is designed to be engaging and educational, connecting classroom concepts to real-world atmospheric phenomena. By conducting these experiments, students will gain a deeper appreciation for the intricate systems that make up our planet's atmosphere and their role in sustaining life on Earth.

Senior High Students: prompt: *in 1000 words or less, select 5 mathematical concepts and 5 corresponding lesson plans*

Here are five (5) mathematical concepts suitable for senior school students, along with corresponding lesson plans that provide a structured approach to teaching these topics.

Algebra: Quadratic Equations

Lesson Plan: Objective: Understand and solve quadratic equations by factorization, completing the square, and the quadratic formula.

- **Introduction (15 minutes):** Begin with a review of polynomial expressions, specifically focusing on the structure of quadratic equations ($ax^2 + bx + c = 0$). Discuss the significance of quadratic equations in various real-life scenarios (e.g., projectile motion).
- **Activity 1 (20 minutes):** Teach students how to solve quadratic equations by factorization. Provide several examples and walk them through the process step by step.
- **Activity 2 (25 minutes):** Introduce the method of completing the square. Use visual aids or diagrams to illustrate how the square is formed and solved. Assign practice problems.
- **Activity 3 (20 minutes):** Teach the quadratic formula and explain the discriminant ($b^2 - 4ac$). Show how the discriminant determines the nature of the roots (real and distinct, real and equal, or complex).
- **Practice and Application (15 minutes):** Assign a mix of problems that require using different methods. Encourage students to choose the most efficient method based on the problem structure.
- **Conclusion (5 minutes):** Recap the three methods and when to use each. Provide a few challenging problems for homework that integrate multiple methods.

Calculus: Differentiation

Lesson Plan: Objective: Understand the concept of differentiation and apply it to solve problems involving rates of change and tangents to curves.

- **Introduction (10 minutes):** Start with a discussion on the concept of limits and how they lead to the derivative. Use a simple example, like finding the slope of a tangent line to a curve at a point.
- **Activity 1 (20 minutes):** Introduce the basic rules of differentiation (e.g., power rule, product rule, quotient rule). Provide examples to illustrate each rule.
- **Activity 2 (20 minutes):** Discuss the physical interpretation of the derivative as a rate of change. Relate it to real-world scenarios like velocity and acceleration.
- **Activity 3 (25 minutes):** Teach how to differentiate common functions (e.g., polynomials, trigonometric, exponential functions). Give students practice problems that include these functions.
- **Application (15 minutes):** Present problems that involve finding the equation of the tangent line to a curve at a given point. Discuss the significance of these problems in fields like physics and engineering.
- **Conclusion (10 minutes):** Summarize the key points and rules of differentiation. Assign homework problems that reinforce the day's lesson, focusing on different applications of differentiation.

Statistics: Probability Distributions

Lesson Plan:

- **Objective:** Understand and analyze different types of probability distributions, including binomial, normal, and Poisson distributions.

- **Introduction (10 minutes):** Review basic probability concepts, then introduce the concept of a probability distribution. Discuss the significance of probability distributions in modeling real-world phenomena.
- **Activity 1 (20 minutes):** Introduce the binomial distribution, its properties, and its applications. Provide examples, such as the probability of getting a certain number of heads when flipping a coin multiple times.
- **Activity 2 (20 minutes):** Explain the normal distribution and its key characteristics, such as the bell curve and standard deviation. Use visual aids to show how data tends to cluster around the mean.
- **Activity 3 (20 minutes):** Introduce the Poisson distribution, discussing its applications in fields like telecommunications and traffic engineering. Provide examples of how it models rare events.
- **Practice (20 minutes):** Give students problems involving each distribution type. Include questions that require calculating probabilities and understanding the distribution's characteristics.
- **Conclusion (10 minutes):** Recap the different distributions covered and their applications. Assign a project where students collect data and identify which probability distribution best models their data.

Geometry: Trigonometry

Lesson Plan:

- **Objective:** Understand and apply trigonometric ratios and identities to solve problems involving right and non-right triangles.
- **Introduction (10 minutes):** Review basic concepts of right triangles and the Pythagorean theorem. Introduce trigonometric ratios (sine, cosine, and tangent).
- **Activity 1 (20 minutes):** Teach students how to use trigonometric ratios to find unknown sides and angles in right triangles. Provide several practice problems.
- **Activity 2 (25 minutes):** Introduce the sine and cosine rules for solving non-right triangles. Explain when to use each rule and provide examples.
- **Activity 3 (20 minutes):** Discuss trigonometric identities (e.g., Pythagorean identity, angle sum and difference identities). Show how these identities can simplify complex trigonometric expressions.
- **Application (15 minutes):** Assign problems that require students to apply trigonometric ratios and identities to solve real-world problems, such as finding the height of a building using an angle of elevation.
- **Conclusion (10 minutes):** Summarize the key concepts and rules. Provide homework that includes a mix of right and non-right triangle problems, along with identity simplifications.

Number Theory: Modular Arithmetic

Lesson Plan:

- **Objective:** Understand the concept of modular arithmetic and its applications in cryptography and computer science.
- **Introduction (10 minutes):** Introduce the concept of congruences and the modulus operator. Use simple examples like clock arithmetic to illustrate the idea.
- **Activity 1 (20 minutes):** Teach students how to perform basic operations in modular arithmetic (addition, subtraction, multiplication). Provide examples and practice problems.
- **Activity 2 (20 minutes):** Introduce the concept of multiplicative inverses in modular arithmetic. Discuss their importance in solving congruences and in cryptography.
- **Activity 3 (20 minutes):** Explain how modular arithmetic is used in cryptographic algorithms like RSA. Discuss the importance of prime numbers in these algorithms.
- **Practice (20 minutes):** Provide problems that involve solving linear congruences, finding multiplicative inverses, and applying modular arithmetic to simple encryption problems.
- **Conclusion (10 minutes):** Recap the key concepts and their applications, particularly in technology and security. Assign a project where students research and present on a cryptographic method that uses modular arithmetic.

For Grade One: Hands-On Activity: Classify Me!

GOAL: Explore animal classification

YouTube search & share: ANIMAL *Groups | Animal Classification for kids/ Miss Ellis* (**Note:** activity includes insects which are not in the video)

Materials: print and cut up the images as single pics, make sets (make as many as required, include six (6) bold heading cards for each set)

prompt: *provide three (3) separate images: a dog, a lion, and a dolphin, each with no background and no text* (**Note:** Repeat each trio of animals listed below)

Instructions:

1. Introduce the topic by writing TOY, FOOD and COLOR on the board. Ask students for items that fit each heading. Write them down. Explain that this is an example of classifying things and today they will learn about classifying some living things.
2. Share the YouTube video. Discuss. On 2nd viewing, pause for students to take notes.
3. In small groups, provide a set of images. Place heading cards across the desk and the images face down. Students rotate choosing an image and putting it in the proper category. Classmates agree or disagree.
4. Match groups. Students compare their sorting.
5. Go over as a class.

Categories for Sorting:

1. **Mammals** (dog, lion, dolphin)
2. **Birds** (parrot, pelican, sparrow)
3. **Reptiles** (snake, turtle, crocodile)
4. **Amphibians** (frog, salamander, toad)
5. **Fish** (goldfish, shark, salmon)
6. **Insects** (butterfly, spider, mosquito)

For Grade 2: Hands-On Activity: Block Based What?

GOAL: Review sequencing and introduce block-based coding

YouTube search & share: 1. *Sequencing | Coding & Computer Science Song/ Scratch Garden* 2. *BLOCK PROGRAMMING for Kids Episode 1 Go Forward/ Smile and Learn*

3. *BLOCK PROGRAMMING for Kids Episode 2 Forward and Right/ Smile and Learn*

4. *BLOCK PROGRAMMING for Kids Episode 3 Forward and Left/ Smile and Learn*

Materials: paper, pen, access to: **https://www.scratchjr.org** or **https://www.kodable.com**

prompt: *for grade 2 level, create a printable workshop activity and write a mixed up, 12 sentence story about getting ready for school that students must arrange in order AND provide an answer key,*

Instructions:

1. Review the need for clear directions when describing how to do something.

2. In pairs, conduct the **prompted activity**.
3. Pair pairs to compare answers. Go over as a class.
4. Turn attention to being clear when telling a computer what to do. The sequencing is critical.
5. Share the 1st YouTube video. Discuss.
6. Discuss block-based coding then share the 2nd YouTube video. Discuss
7. Lead the class in getting on-line, opening your chosen app, and exploring the site.
8. Walk students through an exercise as a class. After that, teacher guided, web-based, self-learning begins. I suggest having students work in pairs.

STEM Literacy Chat: **prompt**: *for grade 2 level, create a printable 200-word conversation between a computer programmer and a student recently introduced to basic block-based coding*

For Grade 4: Hands-On Activity: Matter Relay

GOAL: Explore states of matter, properties of solids, liquids, and gases

YouTube search & share: *What Is Matter? - The Dr. Binocs Show | Best Learning Videos For Kids | Peekaboo Kidz*

Materials: 3 markers, put solid, liquid gas as headings on each of three (3) pieces of paper

prompt: *provide a list of the states of matter - solids, liquids and gases - with 10 qualities identified for each matter listed AND on a separate page provide a list of these 30 total properties in a mixed-up order*

Instructions:

1. Introduce states of matter (Solids, Liquids, and Gases) by asking, *"Who knows what three things make up everything in the world?* "Write a big S, L and G on the board to help out. Discuss.
2. Share the YouTube video. On 2nd viewing, pause to take notes as needed. Discuss.
3. Students write three (3) questions based on the video.
4. In small groups, students ask and answer their questions with classmates.
5. As a class, rotate through groups to share some of their questions with the class.
6. In pairs, conduct the **prompted activity**. Pairs match up and compare answers. Then go over as a class.

Relay Race Time

7. Clear space to make three (3) lanes for a relay race.
8. Place a desk at the end of each lane. Tape a heading sign on each. Place a large book or something between the tables to prevent groups from seeing others' answers.
9. Have three (3) groups of students gather at the other end of each lane.
10. The 1st person in each line gets a marker. Read a description of matter. When done, they walk quickly to the desk up front and place a tick mark under their answer.
11. On returning to their line, they pass the marker to the next person in line. Read another description, and so on until all are completed.
12. Calculate the number correct for each group and declare the winner!!

STEM Literacy Chat: **prompt**: *for grade 4 level, create a printable 250-word conversation between a father and his daughter about what she learned in school today about properties of matter - solid, liquid and gas*

For Grade: Hands-On Activity: Sound Travelers

GOAL: Explore how sound travels through different mediums

Share YouTube search & share: *How Sound Travels Across Different Mediums/ knowledge platform*

Materials: (per group) 3 clear plastic or glass cups (one filled with water, one filled with a solid material like rice or sand, and one empty), tuning fork (or a small bell if a tuning fork is unavailable), metal spoon, long wooden stick (like a dowel), stopwatch or timer, worksheet to record observations

Instructions:

Sound in Gas

1. Hold the tuning fork or small bell in the air, hit it with a spoon to make a sound, and listen.
2. Ask students to observe and describe the sound they hear.
3. Explain that air is a gas and that sound travels through air to reach their ears.
4. Students note their observations on their worksheets.

Sound in Water

1. Fill one cup with water.
2. Strike the tuning fork or bell to create sound, then quickly place the vibrating fork or bell into the water without fully submerging the handle.
3. Ask students close their eyes and observe the sound. They may notice the sound is muffled or seems different than in the air.
4. Explain that sound can travel through water (liquid) but behaves differently than in air.
5. Students note their observations on their worksheets.

Sound in Solid

1. Strike the tuning fork or bell and place the handle against a solid surface like a wooden table or stick.
2. Ask students to press their ear against the table or hold the wooden stick to their ear to listen.
3. Explain that solids allow sound to travel faster and sometimes more clearly than through air or water.
4. Record observations on the worksheet.

Discuss: "Why does sound travel differently through air, water, and wood?" "Which medium carried the sound best?" "Can you think of a real-world example where sound travels through a solid, liquid, or gas?

For Grade 5: Hands-On Activity: Flow With It!

GOAL: Explore algorithms and flowcharts

YouTube search & share: *Algorithms And Flowcharts || Class - 5 Computer || CBSE / CAIE || ICT Education ||Orchids eLearning*

Materials: poster paper, markers, computers or tablets

prompt: *for grade 5 level, create a printable worksheet activity about common flowchart symbols and explain their use AND provide an appropriate activity AND a separate answer sheet with explanations*

Instructions:

1. Introduce the concepts of algorithms and flowcharts.
2. Share the YouTube video. Discuss. On 2nd viewing, pause for note taking.
3. In pairs, students compare their notes.
4. Conduct the **prompted activity**. Go over as a class.
5. As a class, select a topic and create a 7-9 step flow chart flow chart.
6. In small groups, provide materials and groups create a flow chart.
7. As a class, groups present their work.

For Grade 6: Hands-On Activity: Getting Smarter Day by Day!

GOAL: Review coding concepts and build a game

Materials: computers or tablets with access to a block-based coding platform, **prompt**: *for grade 6, create a printable worksheet activity that guides students through coding a game by setting the scene* *then using 4 loops, 4 conditionals, and 4 variables AND a separate answer key*

Instructions:

1. Review the process by writing "*loops, conditionals and variables*" on the board. Discuss.
2. In pairs, conduct the **prompted activity**.
3. Pair pairs to compare their answers.
4. As a class, volunteers share their work.
5. Guide students to your chosen site and ensure they are all on the same page.
6. In pairs, students build a game.
7. Students share their work with other pairs.
8. As a class, discuss the experience.

9

PREPARING FOR THE FUTURE OF AI IN EDUCATION

This chapter shares insights on the evolving world of AI in education and how they impact both education and the new career possibilities your students will encounter.

9.1 EMERGING AI TECHNOLOGIES AND THEIR POTENTIAL IMPACT ON EDUCATION

Quantum computing, *affective* computing, and *augmented reality* (AR) - terms that might sound like they belong in a sci-fi movie - are increasingly becoming a part of the educational lexicon. These technological advancements have the potential to significantly transform how you teach and how your students learn.

In the briefest of terms:

Quantum computing processes vast amounts of data at speeds unattainable by earlier computing capacities.

Affective computing identifies the development of systems and devices that can recognize, interpret, and process human emotions.

Augmented Reality (AR). layers computer-generated enhancements atop an existing reality so students can interact accordingly. The ability to interact with AR can:

- bring textbooks to life,
- turn flat diagrams into interactive 3-D models that students can explore from every angle,
- dissect a frog; or,
- explore the solar system.

AR is the ski goggle type gadget you wear. It makes learning immersive and interactive and helps to clarify complex subjects in highly memorable ways.

Implications for Teaching and Learning

These emerging technologies are set to redefine educational paradigms. *Quantum* computing potentially solves problems too complex for current computers, such as simulating molecular structures in chemistry classes, or optimizing complex logistics in geography lessons.

This could lead to more advanced science, technology, engineering, and mathematics (STEM) curricula that are more aligned with real-world applications.

Affective computing is set to revolutionize the way educators understand and respond to students' emotional and cognitive states. By providing real-time feedback on students' emotional engagement, these systems will help teachers tailor their instructional strategies to better meet individual student needs.

Augmented reality is set to transform abstract concepts into advanced and tangible student experiences in ways traditional learning methods can never do. Whether it's visualizing historical events or solving complex mathematical equations in 3D space, AR caters to different learning styles, increasing comprehension and retention rates.

Challenges and Opportunities

And, as is always the case, nothing comes without challenges. One major concern is the ethical implications surrounding affective computing. How do we ensure AR is used responsibly, especially when they're capable of gathering sensitive emotional data about students?

The digital divide is another concern. It's crucial that, as these technologies develop, they do so in a way that is accessible to all students, regardless of their socio-economic background.

As we navigate these exciting but complex waters, staying informed and critically engaged with the latest AI developments will be key. By understanding both the capabilities and the risks of these emerging technologies, we can better prepare to integrate them into our teaching in ways that are ethical, equitable, and transformative.

9.2 COLLABORATION BETWEEN HUMANS AND AI

Teaching students how to collaborate effectively with AI systems is important. This collaboration extends beyond merely using AI tools; it involves:

- understanding AI capabilities and limitations,
- developing teamwork skills that leverage AI contributions; and,
- navigating the interpersonal dynamics that this new form of teamwork requires.

Projects that simulate real-world AI applications can be a great starting point. For example, conduct a class project using an AI tool to analyze environmental data to predict pollution levels. Have students work in teams to interpret the data and propose solutions based on AI insights. This teaches them about AI's capabilities and how to integrate AI assistance into human-driven, problem-solving processes.

Teamwork skills in an AI-integrated environment also need to evolve. Teach students how to:

- manage AI input,
- integrate it with human insights; and,
- dispute AI findings.

Role-playing exercises are an excellent learning format.

1st: select a topic,

2nd: divide into two teams and provide each team a particular angle surrounding the given topic,

3rd: using AI, teams gather information addressing their particular angle; and,

4th: let the role plays/ discussions begin!

Note: ***AI is not infallible*** *- it can make grammatical and informational errors and produce biased output. Educating students about these limitations helps prevent over-reliance on AI and teaches them the importance*

of critical oversight. Discussions about AI failures, perhaps in tech or ethics classes, provide practical insights into the importance of maintaining a critical eye on AI inputs.

Preparing students for a future where AI is a constant collaborator involves:

a. technical training,
b. soft skills like communication,
c. adaptability: and,
d. ethical reasoning.

Literacy and fluency in manifesting these skills engenders success. (**See section:** 6.3.2)

With AI technologies swiftly evolving, your role as an educator must adopt the mindset of a perpetual learner. As the powers of AI integrate throughout our society, please know that it will eventually reach an equilibrium.

Until that time, we must stay abreast of how these changes impact our role as educators.

Remember that AI education is a continuous journey, not a destination. By continually updating your knowledge and skills, you are keeping up with technological advances. You are also leading by example-showing your students the value of perpetual curiosity and adaptability in a tech-driven world.

9.3 PREPARING STUDENTS FOR AI-INTENSIVE CAREERS

In today's rapidly evolving job market, careers in AI and related fields are becoming essential components of the global economy. As educators, guiding students toward these cutting-edge careers with a clear understanding of AI's unfolding opportunities is very important. Helping students navigate this landscape begins with exposure.

Integrate AI-related content into every subject, especially **STEM** topics. Encourage student participation in AI clubs, hackathons, and summer camps that provide practical experience and networking opportunities. These things make AI impacted career fields more accessible and less intimidating.

For students who show a keen interest, introduce them to specialized courses or programs that focus on AI skills such as coding, statistics, and ethical decision-making. Many universities now offer AI-focused degrees.

Numerous online platforms provide certifications that can bolster a student's skill set even before they step one foot into college. This blend of academic and practical learning ensures students are prepared to enter the AI field and excel.

As the pathways to entering AI careers increase, stay informed. Students will need your educational guidance to successfully navigate the exciting and ever-changing landscape of AI careers.

A FINAL COMMENT

Congratulations! You have completed our deep dive into the dynamic and evolving landscape of AI in education. I hope that the insights, information, strategies, and practical advice shared have sparked your curiosity and equip you with sufficient confidence to harness AI in ways that enrich both your teaching and your students' learning experiences.

At its' core, **Power UP Your Classroom With AI** is about learning how to actively participate in shaping an ethical and educational environment that is innovative, inclusive, and immensely impactful. By integrating AI into your teaching practices, you are enhancing educational outcomes and preparing your students for a future where AI is an integral part of life.

Every educator starts somewhere, and every small step is a leap towards mastering AI in education. Whether it's drafting your first AI-generated activity or setting up interactive group projects, the key is to start at a comfortable pace, learn continuously, and stay open to experimenting with new tools and ideas.

I encourage you to revisit the chapters that most resonate with and continuously seek out new knowledge and resources. Staying informed is the best strategy to guarantee that you and your students remain at the cutting edge of educational possibilities.

Lastly, be open to sharing your experiences and insights with others. Whether it's through online forums, professional development workshops, or casual conversations in the staff room, every shared experience enriches the community's collective knowledge and fosters a more collaborative approach to AI in education.

And with that, we come to both an end, and the beginning.

Thank you deeply for selecting **Power UP Your Classroom With AI**. We covered a lot of ground, including:

Scan the QR code to leave a review.

- AI origins - because knowing and sharing context with students is important,
- Ethical concerns and the anxiety arising from AI use,
- Integration of AI into lesson plans; and,
- Resources for professional development and community building.

Part Two turned our attention to the hands-on aspect of AI in education:

- Resource links by subject and grade level,
- Effective AI use for activity creation,
- Sample activities to inspire; and,
- A brief look toward our AI future.

I wish you great success!
Have fun and happy teaching!

Made in the USA
Monee, IL
04 November 2026

67382324R00098